Insights Into the Heterogeneous Catalysis for the Synthesis of Sugar Alcohols

Insights Into the Heterogeneous Catalysis for the Synthesis of Sugar Alcohols

Anup Prakash Tathod

ELIVA PRESS

Published by Eliva Press
Email: info@elivapress.com
Website: www.elivapress.com

ISBN: 978-1-63648-264-4

© Eliva Press, 2021
© Anup Prakash Tathod
Cover Design: Eliva Press
Cover Image: Freepik Premium

Insights into the heterogeneous catalysis for

the synthesis of sugar alcohols

By Dr. Anup P. Tathod

Ph.D. (Chemical Sciences)

Table of Content   **Page no.**

About Author: *Dr. Anup P. Tathod has done Ph.D. in Chemical Sciences (Catalysis) from CSIR-National Chemical Laboratory, Pune (India) and has worked at Technion, Israel Institute of Technology, Israel as a post-doctoral research fellow. Currently, he is working as a scientist at CSIR-Indian Institute of Petroleum, Dehradun (India). His work interests are; Heterogeneous catalysis, Multifunctional catalysts synthesis and applications, CO₂ utilization. Biomass conversion.*

1.Biomass

Living organisms are being evolved since their origin on the earth. These living organisms are potential resource of energy and often called as biomass. Biomass is the biological material derived from living organism. More specifically it can be defined as "the total mass of living or dead organisms in given area".[1] Chemically biomass is the carbon based material and consists of various organic compounds. These organic compounds contain hydrogen, oxygen and nitrogen in addition to carbon. Alkali metals, alkaline earth metals and some heavy metals are very often found in biomass as a part of important biological molecules. Plants can synthesize carbohydrates from CO_2 and H_2O in the presence of sunlight by the process of photosynthesis. Synthesized carbohydrates are stored in the plant body and further eaten by animals. If plant is not eaten by animal, it undergoes microbial decomposition and such decomposition releases carbon back to the atmosphere. If plant is eaten by animal, after death of the animal its body is decomposed by microorganisms which again returns carbon to the atmosphere. This is called as carbon cycle in which plants are the primary source of energy. Scientific world is looking towards biomass as a renewable resource for the synthesis of chemicals and fuels. Since last few decades biomass has attracted the attention of researchers. Various studies to ascertain the chemical composition, physicochemical properties and applications of biomass and biomass derived compounds in different fields are ongoing.

Depending upon the origin, biomass can be classified into two main categories i.e. animal derived and plant derived. Animal derived biomass includes animal residues and various products obtained from them. Chitin is the most abundant animal derived biomass and second most abundant amongst animal and plant derived biomass. Plant derived biomass consists of plant residues and its products which can further be classified as edible and non-edible biomass. Non edible plant derived biomass commonly termed as lignocellulosic biomass which consists of cellulose, hemicellulose and lignin as main components. These components are bound to each other by various covalent and non-covalent linkages. Cellulose is the most abundant polysaccharide and comprises ca. 50% of lignocellulosic biomass. Cellulose is the homo-polysaccharide of linearly arranged D-glucose monosaccharide units linked together by β-$(1{\rightarrow}4)$ glycosidic linkages (Figure 1.A).[2]

Cellulose is crystalline in nature due to the presence of intra-molecular and inter-molecular hydrogen bonds. These β-(1→4) glycosidic linkages in cellulose make it indigestible to humans, rather cellulose cannot be digested by any vertebrate directly. Symbiotic bacteria in the digestive track of the herbaceous animals possess the enzyme to digest the cellulose. Although cellulose is non-edible to humans, it finds many applications in our day-to-day life e.g. in paper and garment industry.

Hemicelluloses are the second most abundant plant derived polysaccharides and comprise ca. 30% of lignocellulosic biomass. Unlike cellulose, hemicelluloses can be hetero-polysaccharide or homo-polysaccharide made-up of various C5 and C6 sugars depending upon the plant in which it occurs. According to the composition, hemicelluloses are named as xylan (polysaccharide of xylose), arabinan (polysaccharide of arabinose), glucomannan (polysaccharide of glucose and mannose), arabinogalactan (polysaccharide of arabinose and galactose), etc. Xylan present in oat spelt, beech wood, etc. contains some amount of other monosaccharides like arabinose and glucuronic acid. Oat spelt xylan (more precisely arabinoglucuronoxylan) has linear chain of β-(1→4) linked xylose with α-(1→2) bonded 4-O-methyl-D-glucuronic acid unit and α-(1→3) bonded L-arabinose unit (Figure 1.3).[3] Pure xylan (homopolymer of xylose) occurs only in oceanic algae. Generally, hardwood hemicellulose is mainly composed of xylan while softwood hemicellulose contains glucomannan as major component. Arabinogalactan is yet another type of hemicellulose mainly made up of arabinose and galactose. Larch wood arabinogalactan has a chain of β-(1→3) linked D-galactose with the branches made up of β-(1→6) linked D-galactose units and L-arabinose units (some D-glucuronic acid units may present), the structure is shown in Figure 1.A. Unlike most of the other hemicelluloses, larch wood arabinogalactan is water soluble.

Lignin is a complex branched polymer of aromatic compounds comprising ca. 20% of lignocellulosic biomass. Lignin conversion has significant potential for the production of fuels and chemicals as discussed several times in earlier literature.[4-7] Some of the common building blocks of lignin are coniferyl alcohol, sinapyl alcohol and coumaryl alcohol (structures are shown in Figure. 1.B). The exact structure of lignin is not very clear since it depends on various factors such as type of plant, location of plant and age of plant.

By Dr. Anup P. Tathod

Moreover, studies have concluded that lignin may undergo some structural changes during its isolation, so the structure and functional groups present in lignin are significantly governed by the method of isolation. Composition and amount of lignin in plants varies from species to species and even in woods from different parts of the same plant. Softwoods are known to contain higher amount of lignin, followed by hardwoods and grasses contains least amount of lignin.

Inulin is a polymer mainly made up of fructose units linked together via β-(2→1) fructosyl-fructose linkages. Glucose unit may present at the starting of the chain but not essentially. Inulin is present in the vegetables and fruits like onion, garlic, chicory (*Cichorium intybus*), wheat, banana, etc. as a storage carbohydrate. Chicory root consists of abundant amount of inulin and is a major feedstock used for the industrial production of inulin. In chicory inulin, number of fructose units linked to a terminal glucose unit varies from 2-70. Inulin from plants has low degree of polymerization (DP 200 maximum) which varies with the plant species and weather conditions. Inulin which is extracted from the roots of plants consists of fructose as a major component and sucrose, glucose and oligosaccharides in small amount. On the other hand, bacterial inulin has high DP (10,000-100000) and is highly branched.[8] As mentioned above there are two types of inulin polymers, with and without terminal glucose unit. Inulins with a terminal glucose unit are called as α-D-glucopyranosyl-[β-D-fructofuranosyl]$_{n-1}$D-fructofuranoside, while those only constitutes fructose are called as β-D-fructopyranosyl-[α-D-fructofuranosyl]$_{n-2}$-D-fructofuranoside.[9] Structure of the inulin with terminal glucose unit is represented in Figure 1.A. As mentioned earlier, inulin is present in the vegetables and fruits which are the part of daily diet and hence its conversion in to chemicals has to compromise with food. Nevertheless, real substrates which are not suitable for human consumption, like rotten wheat and low-quality garlic can be used efficiently as a feedstock for the production of fructose and its derivatives.

By Dr. Anup P. Tathod

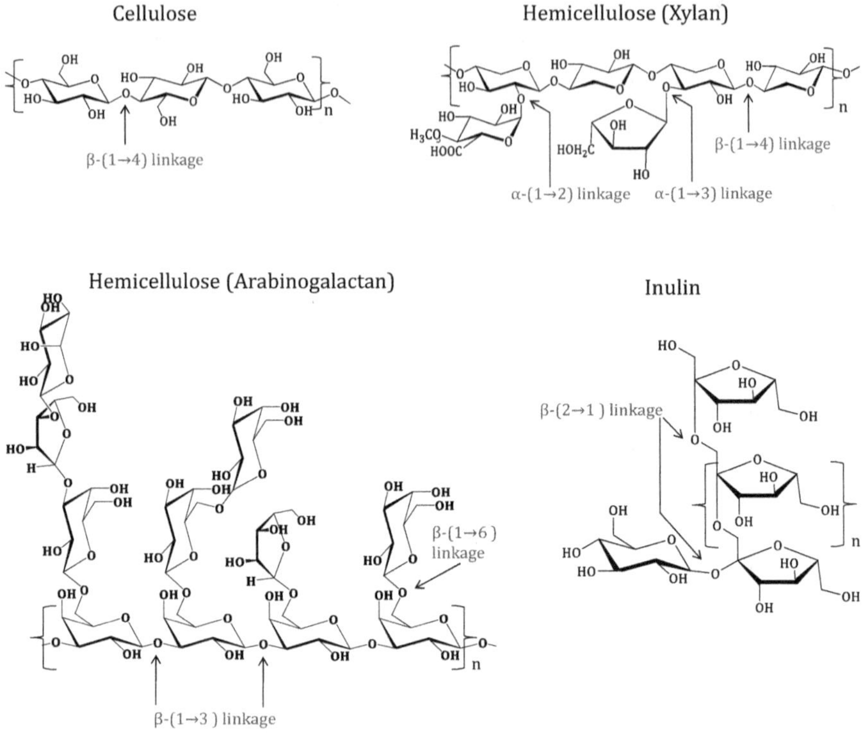

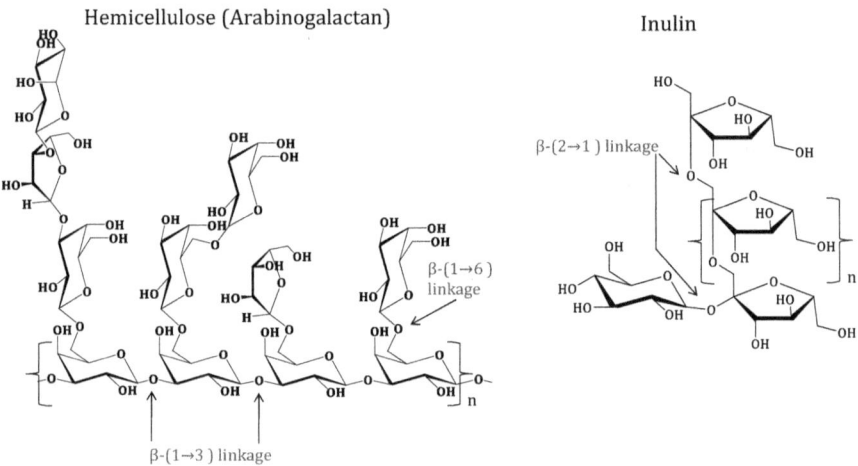

Cellulose

β-(1→4) linkage

Hemicellulose (Xylan)

α-(1→2) linkage   α-(1→3) linkage

β-(1→4) linkage

Hemicellulose (Arabinogalactan)

β-(1→3) linkage

β-(1→6) linkage

Inulin

β-(2→1) linkage

Figure 1A. Structures of poly-saccharides

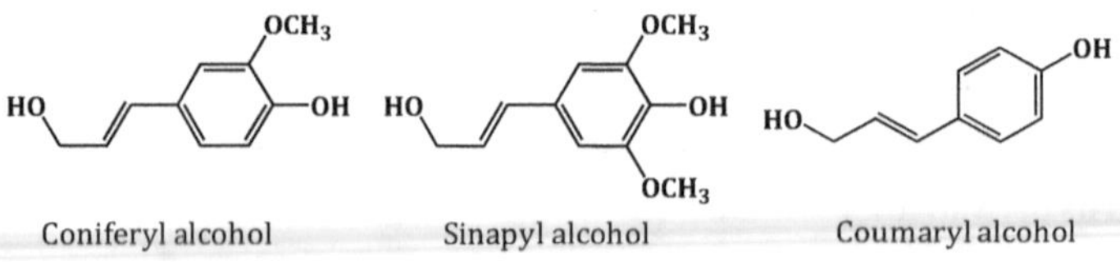

OCH₃

Coniferyl alcohol

Sinapyl alcohol

Coumaryl alcohol

Figure 1.B. Structures of primary building blocks of lignin.

2. Biorefinery

Biorefinery integrates biomass conversion processes to produce fuels and value-added chemicals while petroleum refinery produces fuels and chemicals from fossil feedstock. Petroleum refinery depends on fossil feedstock which is non-renewable and unevenly distributed throughout the world, whereas in biorefinery, renewable feedstock i.e. biomass is used. Hence, concept of biorefinery is the step towards sustainable system for the production of fuels and chemicals. "Sustainable development is the development that meets the needs of present without compromising the ability of future to meet their own needs".[10] Biorefinery deals with the optimum utilization of organic waste which may resolve waste management issue up to some extent. Biorefinery concept is still in the early stage of development and scientific community around the globe is taking efforts for its progress. Though the biorefinery concept is interesting still there are few factors such as raw material availability and its supply, technology suitable to use at commercial level and cost of production which need to be addressed for its success.

As discussed above, biomass has a great potential to replace fossil feedstock at least partially, for the synthesis of fuels and chemicals. Useful chemicals and materials can be obtained from biomass through various conversion processes which can be chemical, thermo-chemical, bio-chemical or physical. Lignocellulosic biomass in the form of agricultural wastes, forest wastes, municipal wastes or agro-based industry wastes can be used as a feedstock in biorefinery for the production of value-added chemicals. Figure 2. represents various types of processes for the conversion of aforementioned wastes in to a variety of products like chemicals, fuels, materials and energy.

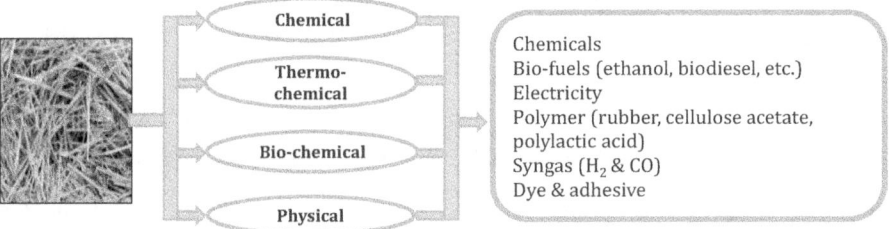

Figure 2. Biomass conversion methods.

By Dr. Anup P. Tathod

Before going into the details about chemicals obtained from the lignocellulosic biomass through various types of conversion processes, it is necessary to be acquainted with these processes. Conversion processes represented in Figure 2. are explained very briefly in next few paragraphs.

Chemical processes: These processes involve chemical conversion of biomass or biomass derived feedstock in to various products. Conversion of lignocelluloses i.e. cellulose, hemicellulose and lignin in to various value-added chemicals is well studied. Although edible feedstock like starch can be used in biorefinery, the discussion in this book is restricted to the conversion lignocellulosic biomass which is non-edible for human being. Homogeneous catalyst like mineral acid (HCl, H_2SO_4) and heterogeneous catalyst like solid acids (metal oxides, amberlyst, sulphonated carbon, heteropolyacids, zeolites), supported metal catalysts (Pt/γ-Al_2O_3, Ru/AC) are often used for the conversion of lignocellulosic biomass in to sugars, furans, sugar acids, sugar alcohols, glycols, etc.

Thermo-chemical processes: These are the processes which not necessarily produce energy but converts biomass in to chemical compounds which can be used for the production of energy more conveniently. Such compounds are more energy dense and have more predictable combustion properties relative to the untreated biomass. For an instance bio-oil, the liquid fuel obtained by subjecting biomass to pyrolysis has higher calorific value (ca. 10,000Kcal/Kg) compared to wood (ca. 3000Kcal/Kg).[11] Under pyrolysis condition, biomass is subjected to high temperatures (350-500°C) in the absence of any oxidant for short period of time to yield pyrolysis oil, generally called as bio-oil. Along with this oil, formation of char and tar is also observed. Syngas can be obtained from biomass by subjecting it to the gasification conditions i.e. to high temperature (700-900°C) under controlled oxygen supply. In this process, biomass is partially combusted to form producer gas and some amount of charcoal. CO_2 and H_2O produced in the partial combustion are further reduced to form CO and H_2. Composition of the gaseous product formed in gasification is ca. 20% H_2, ca. 20% CO, ca. 3% CH_4 and ca. 10% CO_2 (rest is nitrogen).

Bio-chemical processes: These processes involve use of micro-organism for the conversion of biomass in to various products. Anaerobic fermentation of biomass (animal manure and crop waste) for the production of methane rich biogas is one of the oldest known bio-chemical processes. Culture of methanogenic bacteria converts biomass in to biogas (containing ca. 55% of methane) under anaerobic condition. Enzymatic hydrolysis of polysaccharides like cellulose and hemicellulose to yield constituting monomer sugars and fermentation of sugars for the production of ethanol is reported several times in the literature.[12-14]

Physical process: Extraction is the simple method to obtain valuable vegetable oil or seed oil from the biomass. For example; vegetable oil from the plant like Jatropha (*Jatropha curcas*), Soapnut (*Sapindus mukorossi*) are considered as excellent feedstock for the production of bio-diesel. Oils extracted from the plant can be converted in to fuel (bio-diesel) after undergoing base catalyzed trans-esterification reaction.[15] Other than vegetable oils, plants are the source for rubber (rubber tree, *Hevea brasiliensis*), adhesives, dyes (*Indigofera* plant, *Pomegranate* peels), etc.

As mentioned earlier, lignocellulosic material is made up of three main components, cellulose, hemicellulose and lignin, out of which cellulose and hemicelluloses are polysaccharides made up of several monomer sugars (monosaccharides) linked together. Cellulose and hemicelluloses undergo hydrolysis reaction to yield monomer sugars (C6 and C5 sugars) which can further be converted in to various chemicals depending upon the catalyst used and reaction conditions applied. Similarly, many chemical compounds can be obtained from lignin which is made up of various types of aromatic monomers. Amongst several biomass derived chemicals, some are recognized as 'Top value-added chemicals from biomass' by U.S. Department of Energy (in the report 'Top Value Added Chemicals From Biomass, Vol. 1: Results of Screening for Potential Candidates from Sugars and Synthesis Gas, 2004').

Top value-added chemicals from biomass

C3- glycerol, 3-hydroxypropionic acid

C4- succinic acid, fumaric acid, malic acid, aspartic acid, 3-hydroxybutyrolactone

C5- glutamic acid, itaconic acid, levulinic acid, xylitol

C6- 2,5-furan dicorboxylic acid, glucaric acid, sorbitol

Since cellulose is made up of several glucose units linked together, upon hydrolysis it yields glucose. Unlike to cellulose hemicelluloses are made up of various C5 sugars (xylose, arabinose) and C6 sugars (glucose, galactose, mannose), hence upon hydrolysis it gives different types of sugars according to its composition. Homogeneous mineral acids as well as heterogeneous solid acids can be used for the hydrolysis of polysaccharides. Sugars thus obtained, undergoes various reactions to form many useful chemicals. Sugars upon dehydration yields furans i.e. 5-hydroxymethylfurfural from glucose and furfural from xylose while upon hydrogenation C5 and C6 sugar alcohols can be obtained from respective sugars. Furans can further be converted in to levulinic acid, lactic acid, furan dicarboxylic acid, furfuryl alcohols, etc. Other than this, useful chemicals like glycols (ethylene glycol, 1,2-propanediol, glycerol) can also be obtained either from sugars or from sugar alcohols. Some of the sugar derivatives are used as fuel additive, e.g. ethyl levulinate, γ-valerolactone, methyltetrahydrofuran, etc. while others are important platform chemicals. Figure 3. represents the formation of some important chemicals from cellulose and hemicellulose via sugar formation.

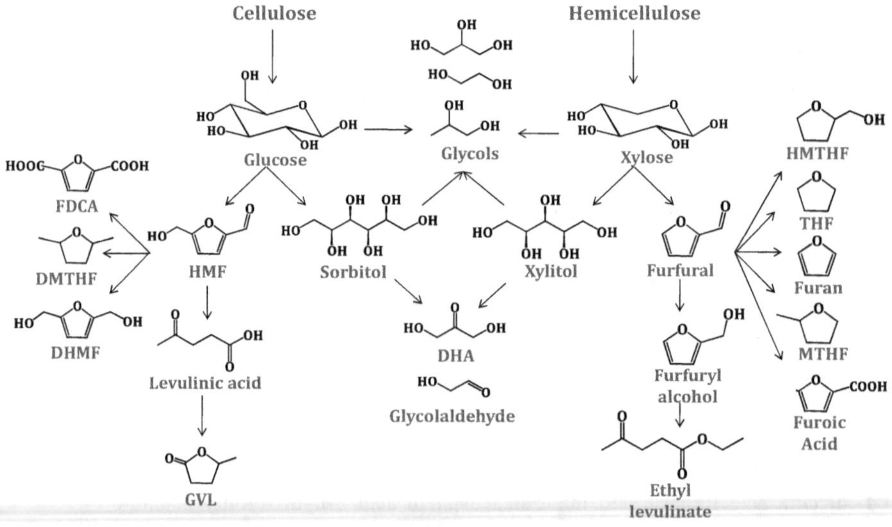

Figure 3. Production of chemicals from cellulose and hemicellulose.

(FDCA: furandicarboxylic acid; DMTHF: dimethyltetrahydrofuran; DHMF: dihydroxymethylfuran; HMF: hydroxymethylfurfural; GLV: γ-valerolactone; DHA: dihydroxyacetone; HMTHF: hydroxymethyltetrahydrofuran; THF: tetrahydrofuran; MTHF: methyltetrahydrofuran).

By Dr. Anup P. Tathod

3. Sugar alcohols

Sugar alcohols are the hydrogenation products of sugars. Sugars are of two types; aldoses (having aldehyde as a functional group) and ketoses (having ketone as a functional group). In the presence of active metal and H_2, aldehyde or ketone group in sugar molecule is hydrogenated to form sugar alcohol. Sugar alcohols are the organic compounds with multiple hydroxyl groups and are also often called as polyols, polyhydric alcohols or polyalcohols. Sugar alcohols have the general formula $HOCH_2(CHOH)_nCH_2OH$ with two hydrogen extra than corresponding sugar (general formula of sugars is $HOCH_2(CHOH)_nCHO$ (aldoses) or $HOCH_2(CHOH)_{n-1}C(O)CH_2OH$ (ketoses). Sugar alcohols are classified according to the length of carbon chain, e.g. tretriol (C4), pentitol (C5), hexitol (C6), etc. Most of the sugar alcohols have their names derived from the name of respective sugars. Xylitol is the sugar alcohol named according to the sugar from which it is derived i.e. xylose. But some sugar alcohols are named differently like sorbitol which is hydrogenation product of glucose. Sugar alcohols have one -OH group attached to each C atom in the chain. The orientation of these -OH groups differentiate one sugar alcohol from other with similar molecular formula. For example, xylitol and arabitol or sorbitol and mannitol are the epimers of each other and do have same molecular formula but different orientation of one -OH group (Figure 4.).

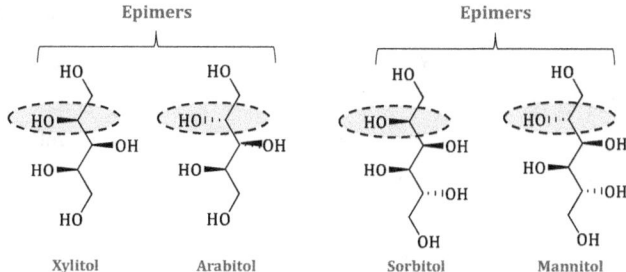

Figure 4. Structures of sugar alcohols.

Sugar alcohols occur naturally in some fruits and berries or can be synthesized from sugars. Sugar alcohols like sorbitol, mannitol and xylitol have some general physicochemical properties as mentioned below.

Properties of sugar alcohols

- Sweet in taste
- High solubility in water
- High thermal stability
- High chemical stability
- Stable at wide range of pH
- Excellent moisture stabilizing effect
- Preserving effect
- High microbial stability
- Excellent compressibility

Because of the aforementioned properties, sugar alcohols find applications in various fields such as pharmaceutical industry, cosmetics and food. DOE (department of energy, U.S.) includes sorbitol and xylitol in the list of 'top value-added chemicals from biomass' which itself reflects their importance in the day-to-day life.

Importance of sugar alcohols

Sugar alcohols find many applications in daily life because of their above discussed characteristic properties. Although, most accredited use of sugar alcohols is as a low-calorie sweetener. Other than this, its importance is well recognized in food industry, pharmaceutical industry, cosmetics and for the production of chemicals. Sorbitol is a C6 sugar alcohol obtained by hydrogenation of glucose which is extensively used for the production of oral hygiene products such as mouth freshener, chewing gums, toothpaste, etc. as well as it is used in humectants. It is used as a precursor for the synthesis of ascorbic acid and out of total sorbitol produced in the world ca. 15% is consumed for ascorbic acid synthesis.[16] Lactic acid can be produced from sorbitol under alkaline hydrothermal condition.[17] It is possible to obtained C3 and C4 polyols from sorbitol by the process of hydrogenolysis..[18] A dehydration product of sorbitol i.e. isosorbide has many applications in the field of medicine, cosmetic and in the synthesis of polymer materials.[19-21] Recently, use of sorbitol for the production of hydrogen is also explored.[22-23] Use of mannitol, (epimer of sorbitol) in medical field is well renowned for its osmotic diuretic properties, as laxative, in the treatment of cerebral oedema and oliguric renal failure.[24] C5 sugar alcohol,

By Dr. Anup P. Tathod

xylitol is well known for its utility as a low calorie sweetener, in cosmetics as well as in pharmaceutical industry.[25-26] Xylitol is used as an ingredient in the preparation of chewing gums, mouth wash and toothpaste very often because of its property to prevent dental decay. [27-29] Chewing of xylitol containing gum is proved to minimize dental caries and its regular use can prevent acute otitis media in childrens.[30] [31] Because of its low calorie content (for xylitol 2.4 cal/g while for sucrose 4.0 cal/g) and insulin independent metabolism, xylitol can be used as a substitute for sugars by diabetics.[32-34] Because of remineralization properties and anti-carcinogenicity, xylitol is important in odontological industries.[35] Applications of sugar alcohols in various fields are summarized in Table 1.

Table 1. Applications of sugar alcohols.

Sr. No.	Sugar alcohol	Applications
1	Sorbitol	Low calorie sweetener, oral hygiene products, humectants, for the production of chemicals and hydrogen.
2	Mannitol	Low calorie sweetener, in the treatment of celebral oedema and oliguric renal failure, as a laxative
3	Xylitol	Low calorie sweetener, having non-diabetes and anti-caries properties, in cosmetics and pharmaceutical industry.
4	Arabitol	Food and pharmaceutical industry

4. Basics of heterogeneous catalysis

Catalytic processes can be classified broadly as homogeneous and heterogeneous catalytic processes based on the nature of catalyst used during the process. Although both homogeneous and heterogeneous catalysts are known for the production of sugar alcohols, the focus of this book is mainly on heterogeneous catalysts. Before we discuss various heterogenous catalytic process for the production of sugar alcohols from various starting materials, let's have a look on the important aspects of the heterogeneous catalysts.

Solid acid: Role of acid in the field of catalysis is well known to scientific society since long time. Importance of the heterogeneous catalyst to develop green methods has been recognized by researchers in last few decades. There are many natural and synthetic solid acids know to catalyze several acid catalyzed reactions. These heterogeneous solid acid catalysts shows two types of acidity i.e. Brønsted and/or Lewis acidity. Type of acidity, amount and strength of acid site and morphological properties like surface area, porosity varies from material to material. Clays have acidic properties and potential to be used as a solid acid catalyst. Montmorillonite-KSF, K-10 and bentonite are some of the clays known for their catalytic properties. Another extensively studied, well known class of solid acid is zeolite. Zeolites are the alumino-silicates having tetrahedral structure of Si and Al surrounded by oxygen atoms. Though the naturally occurring zeolites are microporous, pore size can be controlled in synthetic zeolites. Tetravalent Si bonded with four oxygen atoms is neutral while trivalent Al bonded with four oxygen atoms is negatively charged, hence associated with positively charge counter ions (Na^+). Replacement of these counter ions with protons gives acidity to the zeolites. Typical structure of the zeolite after replacement of counter ions with protons is shown in Figure 5.[36]

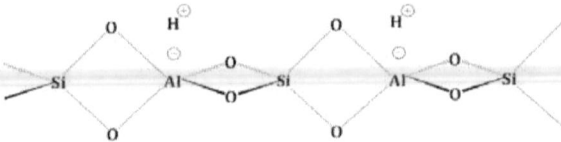

Figure 5. Structure of zeolite.

Metal oxides like silica, alumina, titania, zirconia, etc. and acidified materials like sulphonated silica and carbon in which acidity is introduced by sulphonation reaction, ion exchange resins, etc. shows acidic property and have potential to be used as acidic catalyst.[37-38] Mesoporous material like MCM-41, SBA-15, etc. are of common choice for many acid catalyzed reactions.[39-40] Alumina is often used as a solid acid catalyst or as a support material to prepare supported metal catalysts owing to its thermal stability. Among all forms of alumina, γ-alumina (γ-Al$_2$O$_3$) has wide applications in the field of catalysis. Different forms of alumina can be obtained by subjecting boehmite (γ-AlO(OH)) to thermal treatment at various temperatures. Acidic character of γ-alumina is mainly determined by its surface composition. It shows cubic defect spinel type structure in which Al atoms occupy octahedral and tetrahedral sites while non-spinel sites can be occupied by cations.[41-42] Moreover, γ-Al$_2$O$_3$ shows interesting interaction with water. Lewis acid sites are associated with the Al atom and basic sites are associated with oxide anions at the surface which interact with H$_2$O to form –OH group.[43-45] This interaction involves the adsorption of H$_2$O on the Lewis acid site and subsequent transfer of electron density to the Lewis acid site. After this dissociative adsorption of H$_2$O to modify surface Al coordination with –OH group takes place.[41] γ-Al$_2$O$_3$ upon rehydration loses tetrahedrally coordinated Al from surface and creates hydroxylated octahedrally coordinated Al. Interaction of the γ-Al$_2$O$_3$ with water which results in the formation of hydroxylate Al can be represented as Figure 6.

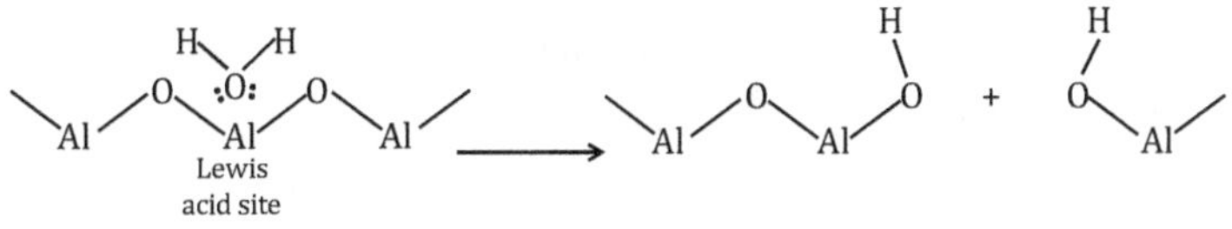

Figure 6. Interaction of alumina with water.

In supported metal catalysts, γ-Al$_2$O$_3$ is used as support material and may show various interactions with metal precursor depending upon the nature of precursor (support-metal interaction is briefly discussed ahead). Density of adsorption sites on the surface is low and hence high dispersion can be achieved but higher metal loading is restricted. Because of these characteristic features γ-Al$_2$O$_3$ is widely used in the preparation of supported metal catalysts. Other than the above-mentioned catalysts there are many more solid acids which could find applications in the field of catalysis. Some of the commonly used solid acids are listed category wise in the Table 2.

Table 2. Examples of solid acids.

Sr. No.	Type	Example
1	Metal oxides	SiO_2, Al_2O_3, ZrO_2
2	Mixed metal oxides	SiO_2-Al_2O_3, WO_3-ZrO_2, Al_2O_3-TiO_2-ZnO
3	Zeolites	H-USY, H-BEA, H-ZSM-5, H-MOR.
4	Heteropoly acids	$Cs_{2.5}H_{0.5}PW_{12}O_{40}$, CsHPW, $H_3PW_{12}O_{40}.6H_2O$
5	Ion exchange resins	Amberlist-15, Nafion-NR50
6	Mesoporous materials	MCM-41, SAB-15,
7	Sulphonated catalysts	Sulfonated-carbon, Sulfonated-ZrO_2, Sulfonated-Silica, Sulfonated-MCM-41
8	Clays	Montmorillonite-KSF, Montmorillonite-K10, Bentonite

Solid base: Solid acid catalysts are well known in the field of catalysis and since last few decades it could succeed in replacing the homogeneous catalysts (mineral acids) in acid catalyzed conversions. Even though solid base catalysts are being used in several industries, extensive catalytic studies on solid base are scanty compared to the solid acid catalysts.[46] Application of homogeneous bases in reactions such as isomerization, addition, alkylation and cyclization is well documented. But the homogeneous nature of catalysts makes their separation and recycling very tricky. Beside this, use of homogeneous bases makes the overall process corrosive and most of the times it needs to be added in the stoichiometric amount. On the other hand, solid bases are easily separable from reaction mixture, safe to handle and can be recycled easily. Analogous to solid acid which has acid sites, solid base has basic sites. In solid acids, acid sites are generated by mixing different kinds of metal oxides but rarely basic sites can be generated by mixing two different kinds metal oxides. In general, basic sites in the form of exposed 'O' atom of metal oxides are generated by removal of H_2O or CO_2 from surface, hence pretreatment at high temperature is needed.[46] Strength of basic sites varies from material to material or the same material may show basic sites with different strength. Strength of the basic site in a particular material is determined mainly by the optimum pretreatment temperature. Strength of the base required for base catalyzed reactions varies from reaction to reaction and depends on abstraction of proton from reactant. As a result, all base catalyzed reactions are not feasible by using single base catalyst. Some of the commonly used solid bases are listed in Table 3. These materials shows some typical properties of base like color change of the acid-base indicator, adsorption of acidic molecules and poisoning of active sites by acidic molecules such as HCl, CO_2, etc.

Table 3. Examples of solid bases.

Sr. No.	Type	Example
1	Metal oxides	CaO, MgO, ZrO_2, SiO-CaO, SiO-MgO, Al_2O_3-MgO (calcined hydrotalcite)
2	Alkali compounds supported on Al_2O_3	KF/Al_2O_3, K_2CO_3/Al_2O_3, $NaOH/Al_2O_3$, KOH/Al_2O_3, etc.
3	Supported alkali metal	Na/Al_2O_3, K/Al_2O_3, K/MgO, Na/zeolite
4	Anion exchangers	Anion exchange resins
5	Zeolites	K, Rb, Cs-exchanged X-zeolite
6	Clay	Sepiolite
7	Phosphates	Natural phosphates, hydroxyapatite

Among all the above-mentioned solid bases, hydrotalcite is one of the most studied and used solid bases. "Hydrotalcites (HT) are the synthetic or natural layered materials made of positively charged two-dimensional sheets of mixed hydroxides with water and exchangeable charge-compensating anions".[47-48] Mg-Al hydrotalcites including their calcination products are very often used as solid base to catalyze various reactions.[49-51] Hydrotalcite can be prepared with various bivalent (Ni, Cu, Co) and trivalent (Fe, Cr) cations other than Mg, Zn and Al. HT exhibit certain characteristic properties which make it appropriate for various base catalyzed reactions. Reactivity of as synthesized HT is governed by the intercalated anion and amount of water present.[49] Two metal hydroxide layers in Mg-Al type HT shows brucite like structure $(Mg(OH)_2)$ in which some bivalent metal cations (Mg^{+2}) are replaced by trivalent metal cations (Al^{+3}). These bivalent metal cations are in close proximity of -OH group. Two brucite layers are placed one over other and held by weak hydrogen bonding. Substitution of bivalent cation by trivalent cation (almost of equal radius) creates excess positive charge which is compensated by intercalating anion.[52] Intercalating compensating anions and water molecules are present in the interlayer space between two brucite layers. HT in as synthesized form does not have activity for most of the base catalyzed reactions because adsorbed water molecules hinder the access of basic sites, hence needs thermal activation typically at 400-500°C for 5-48 h. During heat treatment intercalating anions and water molecules are decomposed and brucite layers are converted into (Mg(Al)O) mixed oxide with higher surface area and porosity.[53-54]

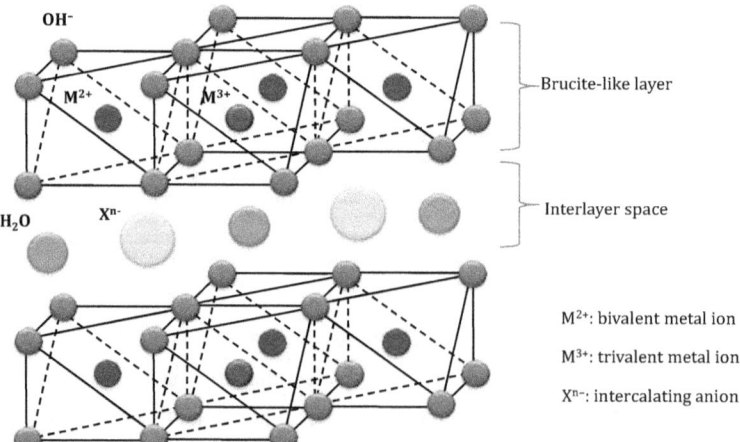

Figure 7. Structure of hydrotalcite.

Moreover the basicity of HT is tunable and fine tuning of basicity is important factor for achieving good catalytic activity.[50] Change in the M^{+3}/M^{+2} ratio or change in the intercalating anions can change the basicity.[55] Basicity of hydrotalcite can be increased by thermal activation. Thermally activated i.e. calcined HT shows good activity in number of reactions. After losing the intercalating anions, O^{-2}-M^{+2} ion pair and strong Lewis basic sites associated with the isolated O^{-2} ions are created which can act as active site for base catalyzed reaction[56-57] and strength of the active site depends on the calcination temperature.[58-59] Calcined HT can be rehydrated to obtained lamellar material (memory effect).[52] Structure of as synthesized HT with two brucite like layers can be represented as Figure 7.

Supported metal catalysts: Supported metal catalysts are the heterogeneous catalysts prepared by anchoring the metal or metals on the suitable support and hence have bi-functional nature. Supported metal catalyst integrates the properties of support as well as the metal. Overall catalytic properties of supported metal catalysts are determined by three factors.

i) Nature of support material: Support materials can be of three types i.e. acidic, basic or neutral. Solid acids, solid bases and neutral materials like carbon are frequently used as support in the preparation of supported metal catalysts (solid acids and solid bases are already discussed in details in earlier sections). Porosity and surface area of support material governs the availability of total binding sites available for anchoring of metal precursor on the support material.

ii) Nature of metal: Metals are being used to catalyze various reactions and ability to catalyze particular reaction is the characteristic property of metal. Neither all the reactions can be catalyzed by a single metal, nor a reaction can be catalyzed by all the metals. Electronic properties of metals are mainly responsible for its catalytic activity towards the reaction. Electronic effect is the result of change in electron density on the metal atom after interaction with support or with other metal.

iii) Support-metal interaction: Support-metal interaction is one of the most important factors which determine the activity of supported metal catalyst. Hence, while choosing a supported metal catalyst to carry out certain reaction, one should consider all above mentioned three aspects. Studies on supported metal catalysts revealed that, various types of interactions may present between supports and metals. Nature of interaction depends on the nature of support and nature of metals as well as on the experimental conditions during its synthesis (pH of solution, temperature of calcination and reduction, etc.) Dispersion of metal is one of the most significant properties of supported metal catalysts. It depends on the support-metal interaction and method of synthesis. Dispersion indicates the number of metal atoms present on the surface out of the total atoms present in the catalyst. Hence, dispersion 'D' can be calculated as;

$$D = N_S/N_T$$

N_S: number of atoms present on the surface

N_T: total number of atoms present.

Smaller the particle size of metal particles higher is the dispersion and specific surface area is inversely proportional to the particle size.[60] Relation between the specific surface area and dispersion can be represented as;

$$S_{sp} = A(N_A/M)D$$

S_{sp}: specific surface area

A: area occupied by an atom on the surface.

N_A: avogadro's number (6.022×10^{23})

M: atomic mass

Other than aforementioned features, supported metal catalysts show effects like spillover effect and steric effect. "Spillover involves the transport of an active species adsorbed or formed on the first phase on to another phase that does not adsorbs or forms the species under same condition". For example, in 'hydrogen spillover' over supported metal catalyst, metal act as donor on which hydrogen species is adsorbed and subsequently migrates to the support which acts as an accepter. Steric effect inhibits the access of reactant molecule to the active site on the catalyst because of bulky groups near the active center. Inhibition of catalytic activity of supported metal catalysts may occur because of adsorption of inhibitory substances, leaching of metal from the support and sintering of metal particle during reaction.

Synthesis methods for supported metal catalysts: It is very important to select the proper method for catalyst synthesis while preparing the supported metal catalyst as it has significant effect on the catalyst properties. Impregnation is one of the most practiced methods for the preparation of supported metal catalysts. In this method solution of metal precursor (active phase) is stirred with support which allows the metal precursor to interact with support material. Two types of impregnation techniques are being used i.e. dry impregnation and wet impregnation. Dry impregnation method is also known as incipient wetness or pore volume impregnation. An adequate amount of precursor solution is used to fill the pores on the supports. Quantity of precursor solution used is just enough to fill the pore volume and not more than that. While in wet impregnation method excess quantity of metal precursor solution is used. In impregnation techniques three processes are involved i.e. transport of metal precursor solution into the pores of support material and diffusion of metal precursor within pores followed by interaction of metal precursor with sites on pore wall. Metal precursors get attached to the support by adsorption (chemical or physical), or by replacement of ligands with surface -OH groups of support. Adsorption of the precursor depends on the surface charge of support. At the pH equals to the point of zero charge, support surface will be neutral, at pH below point of zero charge surface will be positively charged and at pH higher than point of zero charge, surface will be negatively charged. On negatively charged surface, adsorption of positively charged ions will be stronger. Some other methods used for synthesis of supported metal catalysts are homogeneous deposition precipitation, co-precipitation, sol-gel, chemical vapor deposition, etc.[61]

Monometallic and bimetallic catalysts: Monometallic supported metal catalysts are the catalysts which consist of single metal other than support material. As mentioned above, properties of monometallic catalysts are governed by nature of metal, nature of support and support-metal interaction. Bimetallic catalysts consist of two metals other than support material hence the properties of bimetallic catalysts are governed by support-metal interaction and metal-metal interaction along with nature of metals and support. Bimetallic catalyst may exhibit different catalytic properties than individual monometallic catalysts. Generally, second metal in bimetallic catalyst acts as a promoter. "Promoter is the substance that is added to a catalyst in small amounts in order to improve its properties such as activity, selectivity or stability."[62] It is interesting to study the properties of bimetallic catalyst and effects of promoter on the catalytic behavior. Addition of second metal (promoter) may have various effects like, increase in the catalyst activity, better selectivity for particular product or it may improve the catalyst stability. Promoter may cause significant effect because of various interactions or phenomena out of which some are mentioned below;[63]

i) Hydrogen spillover effect: Activated hydrogen species can be transferred from promoter to the accepter metal which may help for the reduction of latter.

ii) Geometric effect: This effect is caused by alteration in the geometry of catalytically active sites due to addition of promoter. Such addition may change the size of 'ensemble' significantly. If adsorbed substrate molecule is in close contact with one particular surface atom, then adsorption site is associated with single atom. If adsorbed substrate molecule is in close contact with more than one atom, then adsorption site is associated with more than one atom or 'ensemble' of atoms. If the size of ensemble is decreased after addition of promoter, then the reaction which needs the larger ensembles will be inhibited. This type of effect is generally referred as ensemble effect.

iii) Electronic effect: Electron transfer may occur between active metal and promoter, which alter the electronic properties of metal. This eventually leads to the change in the catalytic activity of metal. When the interaction or bonding between metal and promoter is strong, this effect is called as ligand effect. Electronic effect does not necessarily include the significant charge transfer but it implies any type of electronic perturbation because of promoter.

iv) Stabilizing effect: Promoter may inhibit the sintering of metal particles which help to improve the dispersion of metal. Promoter can interact with support and hence can form surface shell because of which mobility of metal particles is restricted.

v) Synergistic effect: In this effect, both metal and promoter participate in chemical bonding with reaction intermediate and help to improve the catalytic activity. Promoter may exhibit any one or more than one promoter effect at a time.

By Dr. Anup P. Tathod

5. Catalytic methods for the synthesis of sugar alcohols

Worldwide production of xylitol is ca. 2.4 x 10^4 TPA (tons per annum)[64] and of sorbitol is 6.5 x 10^5 TPA.[65] Demand for sugar alcohols is continuously increasing because of its aforementioned applications in various fields.[66-67] To fulfill the increasing demand, researchers are taking efforts to develop efficient catalytic methods for the synthesis of sugar alcohols. Some reports for the synthesis of sugar alcohols from mono- and polysaccharides are discussed below. The conversion of polysaccharides (e.g. cellulose, hemicelluloses) into sugar alcohols comprises two steps; hydrolysis of polysaccharides to monosaccharides and hydrogenation of monosaccharides (commonly known as sugars) to sugar alcohols. We will discuss the conversion of mono and polysaccharides into sugar alcohols one by one.

Conversion of monosaccharides into sugar alcohols: There are two types of catalysts being used for the production of sugar alcohols from monosaccharides or polysaccharides (cellulose, hemicellulose, inulin, etc.) viz. i) Homogeneous catalysts and ii) Heterogeneous catalysts. Homogeneous catalysts are the catalysts which operate in the same phase where the reaction occurs. Very often these catalysts are co-dissolved in solvent along with reactant. Homogeneous catalysts like enzymes (microbial conversion or bio-conversion) are known for the conversion of sugars in to sugar alcohols since many years. Conversion of xylose to xylitol by *Saccharomyces cerevisiae* or *Candida* is reported in the literature.[64] Application of micro-organisms like *Corynebacterium, Enterobacter liquefaciens, Mycobacterium smegmatis, Petromyces albertensis, Candida* guilliermondii, *Candida tropical, Saccharomyces cerevisiae* and many more are known for the conversion of xylose to xylitol.[35, 68-73] Similarly, conversion of glucose to sorbitol using *Zymmonas mobilis*, fructose to mannitol by *Candida magnolia*,[74] arabinose to arabitol using *Candida entomaea* and *Pichia guilliermondii* are reported.[75-76] Homogeneous catalysts have advantage of good contact with reactant and hence good activity. But it has some obvious drawbacks e.g. difficult separation from product solution and difficulty in reuse. Moreover, some methods involve use of mineral acid, hence needs post reaction neutralization.

Heterogeneous catalysts have overcome the drawbacks of homogeneous catalysts and have attracted the attention of researchers. Conversion of monosaccharides (C5 and C6

By Dr. Anup P. Tathod

sugars) over heterogeneous catalysts like supported metal catalysts is well known. Raney nickel is used for the production of sugar alcohols (xylitol and sorbitol) on industrial scale.[25-26, 77-79] Deactivation of the catalyst because of leaching of nickel and thus high content of nickel in the sugar alcohol solution are the major drawbacks of the methods.[80-81] Therefore purification of sugar alcohol is needed to make it nickel free and suitable for the use in food and pharmaceutical industry.[81] Conversion of glucose to sorbitol over Ni/SiO$_2$, shows negligible leaching but low activity was observed (maximum 45% glucose conversion with 82% selectivity for sorbitol, at 120°C, 5 h).[82] Fructose conversion over Ni based catalyst gives 50% yield of mannitol and sorbitol.[83] Noble metal (Pt, Ru, Rh, Pd) based catalysts were used for the hydrogenation of xylose and glucose to achieve better yield and selectivity for sugar alcohols.[84-87] Although Ru catalyst showed better activity than Ni catalyst, it undergoes deactivation by poisoning of Ru surface by impurities and sintering of metal particles.[88] Effect of support material (MgO, Al$_2$O$_3$, SiO$_2$, Carbon, TiO$_2$) on the catalytic activity of Ru based catalyst in the conversion of lactose was undertaken. 5wt%Ru/Carbon showed highest activity with 96-98% yield of lactitol, however high metal loading (5wt%Ru), high H$_2$ pressure (50 bar) and catalyst deactivation due to metal leaching are the issues to be resolved.[89] Yet another report on the glucose hydrogenation over Pt supported on the activated carbon cloth reveals that high yield of sorbitol (99%) can be achieved. Although this catalyst has some advantages over typical activated carbon catalyst, like high rate of mass transfer and good recyclability, it requires high H$_2$ pressure (80 bar) and very high metal loading (10wt%Pt) to carry out the conversion.[90] Conversion of xylose to yield 99% yield of xylitol was reported over 1wt%Ru-5wt%NiO/TiO$_2$ and catalyst is observed to be stable at reaction conditions (120°C, 2 h, H$_2$ pressure 55 bar).[66] Although few of the above mentioned catalysts give good yield of sugar alcohols from sugars, activities of these catalysts need to be ascertained for the direct conversion of polysaccharides or agricultural waste in to sugar alcohols.

Conversion of polysaccharides into sugar alcohols: Sugar alcohols can be obtained from polysaccharides by hydrolytic hydrogenation in one pot method. Conversion of polysaccharides (cellulose, hemicellulose, inulin) in to sugar alcohols comprises two steps in cascade manner, i.e. hydrolysis of polysaccharides to obtain monosaccharides and

By Dr. Anup P. Tathod

subsequent hydrogenation of monosaccharides in to sugar alcohols. Thus catalyst should have bi-functional nature (should have active sites for hydrolysis and hydrogenation both). Multistep reaction in single pot has advantages like, reduction in the number of steps to obtain final product and reduction in the operating time. Sorbitol production from cellulose over various homogeneous and heterogeneous catalysts is well known.[8, 91-98] However, relatively less number of reports are available for the conversion of hemicelluloses and inulin in to sugar alcohols. Few reports are available in the literature for the production of xylitol from hemicellulose using mineral acids and enzymes as catalyst. Enzymatic conversion of hemicellulose (xylan) using *Candida guilliermondii, Candida athensensis* SB18, *Candida guilliermondii, E. Coli* strain is reported. [99-103] Recently, application of supported metal catalyst (Ru/C) with the combination of mineral acid (H_2SO_4) for the production of xylitol (83% yield) from hemicellulose (xylan) has been reported.[104] Even so, due to homogeneous system, reutilization of these catalysts becomes complicated. To overcome these problems, conversion of hemicellulose in to corresponding sugar alcohols over supported metal catalysts is evaluated without using any mineral acid. Arabinan from beet fibers has been converted to yield arabitol over Ru/AC.[105] Similarly arabinogalactan was converted in to arabitol and galactitol over Ru/MCM-48.[106] But the low yield of sugar alcohols and poor recyclability were the limitations of these catalytic systems. Metal modified beta zeolites can catalyze hydrolytic hydrogenation of hemicellulose, but yield obtained was very low (ca.15% of sugar alcohols).[107] As mentioned above, hydrolysis of hemicellulose (xylan) to xylose by solid acid catalysts is extensively studied [108-113] and use of supported metal catalysts in the xylose hydrogenation reaction is well documented.[66-67, 78, 80, 114] Nevertheless, an efficient one pot method for the production of sugar alcohols directly from hemicellulose with superior yields needs to be developed.

Inulin is mainly made up of fructose and upon hydrolytic hydrogenation gives mannitol and some amount of sorbitol. Not many reports are available for direct conversion of inulin to mannitol over heterogeneous catalyst. Hydrolysis of inulin to yield fructose over acidic zeolite LZ-M-48[115] and conversion of inulin to mannitol over Ru supported on the acidic carbon (acidity was introduced with ammonium per-oxydisulfate) is known to yield 37-40% yield of mannitol.[116] Agricultural waste like wheat straw and bagasse have been used as a substrate for the production of sugar alcohols, but microbe or/and mineral acids were used as catalyst.[101, 103] Application of solid acid catalysts for the conversion of hemicellulose and agricultural waste to C5 sugars and furfural is known.[113, 117-118] Despite that, reports on the direct one pot conversion of untreated agricultural wastes or raw biomass using heterogeneous catalyst are scanty. With regards to all above reports on the conversion of monosaccharides, polysaccharides and agricultural wastes in to sugar alcohols, the heterogeneous catalyst having good activity and recyclability needs to be developed to overcome the drawbacks associated with known methods.

6. Role of support, metal and promoter

In this section of the book, we will discuss about the effect of support, active metal and promoter in the catalytic conversion of mono- and poly-saccharides into sugar alcohols.[119-121]

Role support & metal: As mentioned in the earlier section, supported metal catalysts are bi-functional in nature, exhibiting properties of the both, support and metal. In the case of poly-saccharide's hydrolytic hydrogenation to produce sugar alcohols, the support material is anticipated to play crucial role in first step of hydrolysis. Generally, the hydrolysis step is acid catalyzed and hence the acidic support is expected to show positive effect in the hydrolysis step. Whereas, hydrogenation of mono-saccharides is assumed to be metal catalyzed reaction having no direct influence of the support material. Nevertheless, in hydrogenation reaction too, support can influence the catalytic activity indirectly by altering the support-metal interaction and consequently the catalytic activity of the active metal. For an instance, metal dispersion is one of the imperial properties of supported metal catalyst which has direct impact on the catalytic activity. In fact, catalyst with higher metal dispersion shows superior catalytic activity compare to its analogous catalyst with lower metal dispersion. The metal dispersion is governed by support-metal interaction and therefore nature of support material shows strong impact on the metal dispersion. We have conducted a systematic study to examine the effect of support in the conversion of sugars to sugar alcohol (hydrogenation reaction). For this, Pt based catalysts were prepared using γ-Al$_2$O$_3$ (AL), SiO$_2$-Al$_2$O$_3$ (SA) and activated carbon (AC) as a support material and evaluated in xylose hydrogenation reactions. It can be concluded from the obtained results that the conversion of xylose and yield of sugar alcohols varies with support material despite of similar metal content. Highest yield and conversion were observed over Pt catalyst when AL was used as a support in comparison with SA and AC. This emphasizes that AL is a better support for preparation of Pt based catalysts, i.e. Pt/AL (28% yield of sugar alcohol), Pt/SA (22%) and Pt/C (15%). Further to confirm the effect of support, other catalysts; Ru/AL and Ru/AC were evaluated for xylose hydrogenation at similar reaction conditions. But this time, Ru/AL showed inferior yield of sugar alcohols (21%) compare to Ru/AC (68%). This implies that activity of the catalyst towards the hydrogenation of xylose

By Dr. Anup P. Tathod

is governed by support-metal pair collectively and not by support or metal individually. The superior activity of Ru/AC can be attributed to the very high dispersion of Ru metal on the carbon support. TEM analysis of the catalysts showed that the particle size of Ru on carbon support was approx. 3 nm while that of Pt on alumina support it was 20-30 nm. As particle size goes on decreasing, dispersion of the metal i.e. number of the metal atoms present on the particle surface goes on increasing which eventually creates more active metal sites for hydrogenation reaction. Hence it is so obvious to get higher yield of sugar alcohols over the catalyst with higher metal dispersion.

Role of promoter: Promoter accelerates the activity of catalyst towards particular reaction in one or various ways. The basics of promoter effect is discussed in previous section. It is well known that activity of the supported metal catalysts can be improved by adding promoter to it. There are many reports in the literature in which metal promoters have been used to improve the activity of the catalysts. Majority of the studies about sugar hydrogenation reveal that with Pt or Ru as a base metal, the selectivity for carbonyl group hydrogenation can be improved with the addition of promoters such as, Sn, Ga (non-transition element) or Fe (transition metal). From most of the studies it is found that among all the promoters, Sn shows better results to obtain unsaturated alcohols from unsaturated aldehydes. We have conducted several experiments to study the promoter effect of Sn for sugar hydrogenation. First of all, effect of Sn as a promoter in Pt and Ru based catalysts was studied for xylose hydrogenation. Effect of support and metals can be more prominent in bimetallic catalysts compare to monometallic catalysts. Because, excluding support-metal interaction, the interactions like support-promoter and metal-promoter interaction can co-exist in bimetallic catalysts. Taking this into account, bimetallic catalysts using Sn as a promoter, Pt or Ru as a metal and AL, SA or AC as a support, were prepared and evaluated for xylose hydrogenation. In the case of alumina and silica-alumina based catalysts Sn showed significant improvement in the catalytic activity in terms of improved conversion and selectivity towards xylitol when added along with Pt. But in case of carbon-based catalyst, no significant improvement after addition of Sn was observed. Moreover, the promoter effect was observed most prominently in case of Pt/AL. Monometallic Pt/AL showed 28% yield of sugar alcohols while bimetallic Pt-Sn/AL showed 73% yield of sugar alcohols. Additionally, formation of furans as a side product was observed over all the catalysts. However, it was more in case of AL and SA based catalysts compare to AC based catalysts owing to the higher acidity of AL and SA compare to AC. Formation of furan from sugar is assumed to be acid catalyzed reaction. Interestingly, formation of furan compounds was suppressed up to much extend (down to 1%), after incorporation of Sn in the Pt/AL catalyst. However, over AC supported catalyst, almost similar yield of furans was observed after incorporation of Sn. Superior yield of sugar alcohols over Pt-Sn/AL catalyst was possibly due to the availability of higher number of active metal sites than monometallic Pt/AL catalyst due to higher metal dispersion in the former. The TEM study reveals that the average particle size of Pt in Pt/AL was 20-30 nm while in Pt-Sn/AL it was 10-20 nm. Besides increase in active sites some other factors were also playing role in enhancing the sugar alcohols formation over Pt-Sn/AL catalyst. It is assumed that Sn interacts with the acidic sites of AL to form a complex and this in turn may decrease the number of acid sites in the catalyst. This was confirmed by the NH_3-TPD study of the Pt/AL and Pt-Sn/AL catalyst. This implies that the overall formation of various products from the xylose is strongly affected by the nature of support. Furthermore, when promoter effect of Sn was explored in Ru based catalysts, inferior yields were obtained over bimetallic Ru-Sn/AL and Ru-Sn/AC catalysts compare to the respective monometallic catalysts. It can be concluded that improved activity of the bimetallic catalyst in together is governed by support, metal and promoter.

Another important factor which governs the activity of bimetallic catalyst is the metal:promoter ratio. The promoter shows positive effect on the catalysis only when it is present in appropriate amount and suitable chemical state. Let's see these two important aspects one by one. The appropriate amount of promoter and the appropriate metal:promoter ratio is the crucial factor while using promoter, otherwise the promoter can act as a poison if used in inappropriate amount. We can better understand this concept with familiar example of Pt-Sn/AL catalyst. For this study, we prepared the series of bimetallic catalysts with Pt loading of 3.5wt% and Sn with varying amount (from 3.5wt% to 0.22wt%) and were subjected to XRD analysis. The study revealed that, catalysts with Sn content higher than 1.75wt% showed the formation of Pt-Sn intermetallic system like (PtSn and Pt_3Sn). On the other hand, the catalysts with lower Sn content did not show the formation of any intermetallic systems. When these bimetallic catalysts were evaluated for xylose hydrogenation, the catalyst with Sn content of 1.75 and 3.5wt% showed inferior yield of sugar alcohols while the catalysts with Sn content of 0.22, 0.43 and 0.87wt% showed superior yields of sugar alcohols. This indicates that the formation of Pt-Sn intermetallic system is affecting the catalytic activity negatively. Hence it is very important to use optimum amount of the promoter. Next factor which has very strong impact on the promoter effect is chemical state of the promoter. This aspect also can be understood with the classic example of Pt-Sn/AL catalytic system. As discussed in the above paragraph, Sn showed promoter effect when its concentration was less (less than 0.87wt%). With increasing amount of Sn in the Pt-Sn/AL catalysts, the activity was observed to be decreased. This is because, Sn can form tin-aluminate type structure with alumina support which creates surface shell on the alumina surface during calcination treatment. Sn in the form of tin-aluminate cannot be reduced easily and remains as Sn (II) or Sn (IV) species even after reduction at higher temperature. However, only limited amount of Sn can form tin-aluminate type structure on the alumina surface. Surplus amount of Sn could not find the proper alumina sites to form tin-aluminate and thus remains as tin-oxide which can be reduced to metallic tin upon reduction. Such a metallic tin can form Pt-Sn intermetallic system with Pt. Hence, if amount of Sn is higher than optimum amount it shows negative impact on the catalytic activity. To study the mechanism by which Sn enhances the catalytic activity of Pt/AL towards hydrogenation of sugars, systematic study was conducted. It is clear that when Sn is present in the form of tin-aluminate, it can act as promoter. But how does it work? Sn forms the surface shells of tin-aluminate on the alumina surface which restrict the mobility of Pt particle during catalyst synthesis and post synthesis treatments. This helps to improve the Pt dispersion which ultimately leads to the improved catalytic activity. The TEM analysis proved that the Pt dispersion was much higher in Pt-Sn/AL bimetallic catalyst compare to Pt/AL monometallic catalyst. Other than improved dispersion, Sn can enhance the activity by the polarizing effect. Sn in the form of tin-aluminate is present as Sn (II) or Sn (IV) species, which can polarize the carbonyl group of the sugar molecules. The polarized C=O bond can easily be hydrogenated in presence of active metal like Pt and hydrogen. At the same time, Sn in the form of metallic tin or Pt-Sn inter metallic systems cannot polarize the C=O bond and hence cannot exhibit promoter effect. XPS study proved that the oxidation state of the Sn in the catalyst with lower Sn

By Dr. Anup P. Tathod

content was +2 or +4 while in the catalyst with higher content of Sn, the formation of metallic tin was observed. These observations help to explain, "why Sn could not show promoter effect when carbon was used as a support?". Unlike to the alumina support, on the carbon support Sn cannot form surface shells and hence nigher higher Pt dispersion nor polarization of C=O bond occurs. After successful improvement in the yield of sugar alcohols from xylose using Pt-Sn/AL catalyst, the catalyst was evaluated for other C5 and C6 sugars. Similar to xylose, Pt-Sn/AL showed superior activity than Pt/AL in the hydrogenation of other sugars like arabinose, glucose, fructose, galactose, etc.

So far, we have discussed the conversion of monosaccharides in to sugar alcohols over various monometallic and bimetallic catalysts. Bimetallic Pt-Sn/AL catalyst showed extraordinarily good results in the hydrogenation of C5 and C6 sugars to improve the yields of sugar alcohols. Nevertheless, the production of monosaccharides from polysaccharides with high selectivity is tedious because acid catalyzed hydrolysis is associated with some other acid catalyzed reactions. Most of the sugars occur naturally in the form of polysaccharides i.e. cellulose, hemicellulose, starch, inulin, etc. (structural properties of these polysaccharides are already explained in previous section). Production of sugar alcohols from polysaccharides is a two-step process including hydrolysis of polysaccharides in to constituting monosaccharides and subsequent hydrogenation of monosaccharides in to sugar alcohols. Production of sugar alcohols from polysaccharides by hydrolytic hydrogenation in a one pot manner is the better alternative as it avoids separation and purification of intermediate chemical compound (sugars) and save time as well as resources. Considering above facts, conversions of hemicelluloses and inulin were undertaken to yield sugar alcohols in one pot fashion. Hemicelluloses are the second most abundant plant derived polysaccharides after cellulose and non-edible to humans; therefore, its conversion to sugar alcohols is one of the best ways to utilize it.

Similar to the monosaccharide conversion, Pt-Sn/AL catalyst showed better performance camper to Pt/AL catalyst in the conversion of polysaccharides. Study was conducted using xylan, arabinogalactan and inulin as a substrate. Regardless the substrate, Pt-Sn/AL showed superior yields of sugar alcohols compare to Pt/AL, due to obvious reasons explained earlier. However, conversions and yields of sugar alcohols were less in

By Dr. Anup P. Tathod

case of poly-saccharides conversion than that of mono-saccharides conversion. In the poly-saccharide conversion, hydrolysis is rate determining step requires higher temperature and longer reaction time. The hydrolysis is acid catalyzed step and in our catalytic system alumina support is the only source for acid sites. Additionally, alumina is not very strong acid and this results into very slow hydrolysis of poly-saccharides which consequently results into lower yields of sugar alcohols. Pt-Sn/AL also showed considerably good results for the production of sugar alcohols from raw agriculture wastes e.g. rice husk, wheat straw and sugarcane bagasse.

By Dr. Anup P. Tathod

7. Effect of solid base as an additive

In previous section we have discussed about the promoter effect of second metal in the hydrogenation of mono-saccharides and hydrolytic hydrogenation of poly-saccharides. The catalytic activity of monometallic or bimetallic catalyst can further be improved using solid base as an additive.[122] For this study, hydrotalcite was used as a solid base. General properties of solid bases and characteristic feature of the hydrotalcite as a solid base is already discussed in details in earlier section.

Hydrotalcite was used as an additive along with Pt/AL or Pt-Sn/AL catalyst in the hydrogenation of xylose. In both cases, it showed positive effect on the catalytic activity. The complete conversion with improved selectivity could be achieved in presence of hydrotalcite at milder reaction conditions. Before understanding the plausible mechanism through which hydrotalcite improves the catalytic activity, we should be aware of the keto-enol tautomerism of the sugar molecules. It is well known that the concentration of sugar in open chain form in neutral medium like water is very small at equilibrium and most of the sugar molecules are in the closed chain form (having hemiacetal linkage). It is known that, in alkaline medium glucose molecule can undergo Lobry de Bruyn-Albeda van Ekenstein transformation to form its isomer fructose. This transformation proceeds through the ring opening of glucose molecule and formation of enolate and enediol ions as intermediates. Similarly, other sugar molecules can also undergo this transformation in alkaline medium. The step involved in the Lobry de Bruyn-Albeda van Ekenstein transformation are shown in Figure 8.

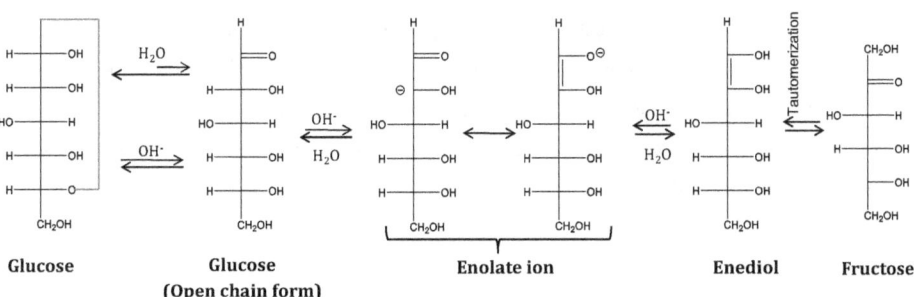

Figure 8. Isomerization of glucose to fructose in alkaline medium.

By Dr. Anup P. Tathod

With the addition of hydrotalcite to the reaction mixture, pH of the reaction mixture becomes alkaline and a greater number of sugar molecules can be in open chain form. The increased number of open chain sugar molecules can be confirmed with the help of UV-Vis adsorption study. Glucose molecule is UV active if it is present in open chain form due to exposure of –CHO group, while absence of –CHO group in cyclic form makes glucose UV inactive. Based on this fact, UV absorption study of glucose was done to confirm that more glucose molecules are in open chain form under alkaline condition. Although, in the alkaline reaction medium, sugar molecules are proved to be in open chain form which further helps for its easy hydrogenation, pH of the reaction medium is crucial factor to determine this effect. We have conducted experiments with several solid bases which results into the variable pH of the reaction medium when added along with Pt-Sn/AL catalysts. The effect of solid bases like, hydroxyapatite, magnesium oxide and calcium oxide were compared with the effect of hydrotalcite as an additive on xylose conversion over Pt-Sn/AL catalyst. Among all the catalytic system evaluated, magnesium oxide and hydrotalcite showed best results, where pH of the reaction mixture was around 9-10. Addition of hydroxyapatite and calcium oxide did not show much improvement where pH of the reaction medium was around 7.5 and 12 respectively. Aforementioned results indicate that optimum yield can be achieved when pH of reaction mixture is 9-10 and start decreasing at pH higher than 10 and lower than 9, because at lower pH (7.5 or lower) less number of sugar molecules is present in open chain form while at higher pH side reactions may prevail.[123]

When the effect of hydrotalcite was examined as an additive along with Pt-Sn/AL catalysts in the conversion of polysaccharides, no improvement in the yield of sugar alcohol was observed. Adversely, in some cases conversion and yields were observed to be decreased after addition of hydrotalcite. This is very obvious, as hydrotalcite is a solid base and hydrolysis of polysaccharides is favored by acid. Thus, the presence of base interferes with the acid site of the catalyst and limits the hydrolysis of polysaccharides.

By Dr. Anup P. Tathod

8. Summary and Conclusions

Sugar alcohols are the vital chemical compounds having applications in various fields such as food and pharmaceutical industries, in oral hygiene products, for the production of chemicals, hydrogen generation and most importantly as a low-calorie sweetener. These sugar alcohols can be synthesized from C5 or C6 sugars upon hydrogenation as well as from polysaccharides by hydrolytic hydrogenation. Cautious study of the literature shows that, although there are many reports for the production of sugar alcohols from monosaccharides (C5 and C6 sugars) are available, very few methods for the conversion of polysaccharides and even less for the one pot conversion of agricultural wastes are known. Moreover, conventional methods are allied with some drawbacks like low yield of sugar alcohols, requisite of strong reaction conditions and poor recyclability due to deactivation of the catalyst. Because of significant applications, demand for sugar alcohols is increasing continuously; hence it is indispensable to develop an efficient method for the synthesis of sugar alcohols. In our research work, conversion of various C5 and C6 sugars (xylose, arabinose, glucose, galactose and fructose), isolated polysaccharides (xylan, arabinogalactan and inulin) and agricultural wastes (bagasse, rice husk and wheat straw) was carried to yield C5 and C6 sugar alcohols (xylitol, arabitol, sorbitol, mannitol and galactitol) using monometallic and bimetallic supported metal catalysts.

Considering the superior catalytic activity in the hydrogenation of sugars, Pt and Ru based supported metal catalysts were evaluated for the hydrogenation of various C5 and C6 sugars. Catalytic activity of the synthesized monometallic and bimetallic catalysts towards the formation of sugar alcohols from monosaccharides was evaluated and correlation between activity and physicochemical properties of the catalyst was established. It was observed that Pt-Sn/AL catalyst (with Pt:Sn ratio equals to 1:0.12 (*wt/wt*)) showed the best results for the production of C5 and C6 sugar alcohols from corresponding sugars. Sn in ionic state, (+2) or (+4) promotes the activity of Pt/AL catalyst towards hydrogenation of sugars. Sn can improve the dispersion of Pt by forming surface shell on the AL and this ultimately increases the number of active metal sites in the catalyst. Consequently, better results for the hydrogenation of sugars can be achieved over Pt-Sn/AL bimetallic catalyst than Pt/AL monometallic catalyst. Furthermore, Sn (II) or Sn (IV) can polarize the C=O bond of the carbonyl group of sugar molecules which can be hydrogenated easily. The yield was observed to be decreased with increasing Sn loading in Pt-Sn/AL catalysts, because formation of Pt$_3$Sn and PtSn intermetallic systems were observed in these catalysts. Nevertheless, yield of sugar alcohols obtained over Pt-Sn/AL bimetallic catalyst with any Pt:Sn ratio was better than monometallic Pt/AL catalyst as in bimetallic catalyst with higher Sn loading too, some amount of Sn was present in (+2) or (+4) state. When the effect of supports, metals, promotes was examined in the conversion of xylose, best results were observed over Pt-Sn/AL bimetallic catalyst compared with any other catalytic system. Sn could not demonstrate promoter effect when activated carbon was used as support, since Sn cannot form complex with carbon and can be reduce to Sn (0), hence formation of PtSn intermetallic system is possible. Similar to xylose, hydrogenation of arabinose, glucose, galactose and fructose was carried over Pt-Sn/AL catalyst and yield of sugar alcohols was observed to be enhanced by more than 2 times compared with Pt/AL catalyst. The Pt-Sn/AL catalyst showed good recyclability in the conversion of C5 and C6 sugars.

Since improved yields of sugar alcohols were obtained over Pt-Sn/AL catalyst compare to Pt/AL catalyst in the conversion of C5 (xylose and arabinose) and C6 (glucose, galactose and fructose) sugars, efficiency of Pt-Sn/AL catalyst was evaluated in the conversion of isolated polysaccharides to yield sugar alcohols. Conversions of isolated polysaccharides (xylan, arabinogalactan and inulin) were carried over Pt-Sn/AL catalyst and reaction conditions were optimized to achieve maximum yield of sugar alcohols. Time required for complete conversion of polysaccharides is much longer than that for conversion of monosaccharides. This is because, conversion of polysaccharides in to sugar alcohols comprises two steps; hydrolysis of polysaccharides to yield monosaccharides and subsequent conversion of monosaccharides in to sugar alcohols. One pot conversion of agricultural wastes in to sugar alcohols is much convenient method as it skips over two steps, e.g. isolation of polysaccharides and hydrolysis of polysaccharides to monosaccharides. Considering this, selective conversion of hemicellulose in the agricultural wastes to produce sugar alcohol (C5 sugar alcohols) was conducted over Pt-Sn/AL catalyst.

By Dr. Anup P. Tathod

When conversion of agricultural wastes was carried out over Pt-Sn/AL catalyst, yield of sugar alcohols was observed to be improved by 1.5 to 2 times compared to Pt/AL. It can be concluded that Pt-Sn/AL catalyst is applicable at wide range of reaction conditions and can be used for the synthesis of sugar alcohols from variety of substrates.

Dramatic improvement in the yield of sugar alcohols was observed over Pt-Sn/AL catalyst compare with Pt/AL catalyst in the conversion of monosaccharides, polysaccharides and agricultural wastes. Promoter effect of Sn in Pt/AL catalyst can be explained with the help of ability of ionic Sn to polarize the C=O bond in sugar molecules. Polarization of C=O bond is possible only if sugar molecule is present in open chain form, hence efforts were taken to generate sugar molecules in open chain form in reaction mixture. It is known that under alkaline condition sugars can undergo isomerization reaction (Lobry de Bruyn-Alberda van Ekenstein transformation) through open chain intermediates (like enolate and enediol). This means more sugar molecules must be present in open chain form in alkaline medium than neutral or acidic medium, which can be hydrogenated if active metal and hydrogen is available. Considering this, effect of solid base, hydrotalcite in the conversion of monosaccharides and polysaccharides over monometallic and bimetallic catalysts was examined. Use of solid base along with supported metal catalyst is a convenient method to improve the yields of sugar alcohols from sugars. Maximum 82% and 68% yield of C5 and C6 sugar alcohols could be achieved from xylose and glucose, respectively using hydrotalcite in combination with Pt/AL monometallic catalyst. Complete conversion of the sugars with better yields of sugar alcohols could be achieved at much milder reaction conditions (60-80°C) within short period of reaction time (4 h), in the presence of hydrotalcite. UV-Vis study of glucose proved that more sugar molecules are in open chain form in alkaline medium than in neutral medium. Sugar molecule in open chain form can be hydrogenated easily at milder reaction conditions. Hence it is possible to achieve complete conversion at low temperature in presence of hydrotalcite (as it makes pH of reaction mixture alkaline). Moreover, at lower temperatures side reactions to yields undesired products like glycols and furans are not prominent; therefore, better selectivity for sugar alcohols could be achieved. Further enhancement in the yield of sugar alcohols over Pt-Sn/AL catalyst in combination with hydrotalcite was because of combined effect of hydrotalcite and Sn in (II) or (IV) state. hydrotalcite help to generate sugar molecule in open chain form and ionic Sn can polarize the C=O bond of sugar molecules (already in open chain form under alkaline condition) which can be easily hydrogenated. Unfortunately, in the conversion of polysaccharides, yields could not be improved by adding hydrotalcite along with Pt-Sn/AL catalyst, because such conversions need higher temperature (110-190°C) and at higher temperatures side reactions to yield glycols and furans are prominent.

It can be concluded that, Sn acts as a promoter in the Pt-Sn/AL bimetallic catalyst and improvement in the yields of sugar alcohols can be achieved over Pt-Sn/AL catalyst in the conversion of C5 and C6 sugars, polysaccharides and agricultural wastes. Further improvement in the yields can be observed from C5 and C6 sugars, if solid base (hydrotalcite) is used as an additive along with the Pt-Sn/AL catalyst.

9. Bibliography

1.   Pasha, J.; Kandagatla, B.; Sen, S.; Seerapu, G. P. K.; Bujji, S.; Haldar, D.; Nanduri, S.; Oruganti, S., Amberlyst-15 catalyzed Povarov reaction of N-arylidene-1H-indazol-6-amines and indoles: a greener approach to the synthesis of exo-1,6,7,7a,12,12a-hexahydroindolo[3,2-c]pyrazolo[3,4-f]quinolines as potential sirtuin inhibitors. *Tetrahedron Letters 56* (18), 2289-2292.
2.   Berchel, M.; Haddad, J.; Le Corre, S. p. S.; Haelters, J.-P.; Jaffrès, P.-A., Synthesis of lipid-based unsymmetrical O,O-dialkylphosphites. *Tetrahedron Letters 56* (18), 2345-2348.
3.   Phadnis, P. P.; Wadawale, A.; Priyadarsini, K. I.; Jain, V. K.; Iwaoka, M., Synthesis, characterization, and structure of trans-3,4-dihydroxy-1-selenolane {DHS(OH)2} substituted derivatives. *Tetrahedron Letters 56* (18), 2293-2296.
4.   Harkin, J., Recent Developments in Lignin Chemistry. In *Naturstoffe*, Springer Berlin Heidelberg: 1996; Vol. 6, pp 101-158.
5.   Amen-Chen, C.; Pakdel, H.; Roy, C., Production of monomeric phenols by thermochemical conversion of biomass: a review. *Bioresource Technology* **2001,** *79* (3), 277-299.
6.   Huber, G. W.; Iborra, S.; Corma, A., Synthesis of Transportation Fuels from Biomass:‰ Chemistry, Catalysts, and Engineering. *Chemical Reviews* **2006,** *106* (9), 4044-4098.
7.   Dorrestijn, E.; Laarhoven, L. J. J.; Arends, I. W. C. E.; Mulder, P., The occurrence and reactivity of phenoxyl linkages in lignin and low rank coal. *Journal of Analytical and Applied Pyrolysis* **2000,** *54* (1–2), 153-192.
8.   Dhepe, P.; Fukuoka, A., Cracking of Cellulose over Supported Metal Catalysts. *Catalysis Surveys from Asia* **2007,** *11* (4), 186-191.
9.   Ronkart, S. b. N.; Blecker, C. S.; Fourmanoir, H. l. n.; Fougnies, C.; Deroanne, C.; Van Herck, J.-C.; Paquot, M., Isolation and identification of inulooligosaccharides resulting from inulin hydrolysis. *Analytica Chimica Acta* **2007,** *604* (1), 81-87.
10.   Doelle, H. W., Biomass and Organic Waste Conversion to Food, Feed, Fuel, Fertilizer, Energy and Commodity Products. *Biotechnology, Ed. Horst W. Doelle* **2003**.
11.   Chang, Y.-M., On pyrolysis of waste tire: Degradation rate and product yields. *Resources, Conservation and Recycling* **1996,** *17* (2), 125-139.
12.   Yang, B.; Dai, Z.; Ding, S.-Y.; Wyman, C. E., Enzymatic hydrolysis of cellulosic biomass. *Biofuels* **2011,** *2* (4), 421-449.
13.   Yoshida, M.; Liu, Y.; Uchida, S.; Kawarada, K.; Ukagami, Y.; Ichinose, H.; Kaneko, S.; Fukuda, K., Effects of Cellulose Crystallinity, Hemicellulose, and Lignin on the Enzymatic Hydrolysis of Miscanthus sinensis to Monosaccharides. *Bioscience, Biotechnology, and Biochemistry* **2008,** *72* (3), 805-810.
14.   Mielenz, J. R.; Doran-Peterson, J.; Jangid, A.; Brandon, S.; DeCrescenzo-Henriksen, E.; Dien, B.; Ingram, L., Simultaneous Saccharification and Fermentation and Partial Saccharification and Co-Fermentation of Lignocellulosic Biomass for Ethanol Production. In *Biofuels*, Humana Press: 2009; Vol. 581, pp 263-280.
15.   Schuchardt, U.; Sercheli, R.; Vargas, R. r. M., Transesterification of vegetable oils: a review. *Journal of the Brazilian Chemical Society* **1998,** *9*, 199-210.
16.   Gallezot, P.; Cerino, P. J.; Blanc, B.; Fleche, G.; Fuertes, P., Glucose hydrogenation on promoted raney-nickel catalysts. *Journal of Catalysis* **1994,** *146* (1), 93-102.
17.   Ramírez-López, C. A.; Ochoa-Gómez, J. R.; Gil-Río, S.; Gómez-Jiménez-Aberasturi, O.; Torrecilla-Soria, J., Chemicals from biomass: synthesis of lactic acid by alkaline hydrothermal conversion of sorbitol. *Journal of Chemical Technology & Biotechnology* **2011,** *86* (6), 867-874.
18.   Chen, X.; Wang, X.; Yao, S.; Mu, X., Hydrogenolysis of biomass-derived sorbitol to glycols and glycerol over Ni-MgO catalysts. *Catalysis Communications* **2013,** *39* (0), 86-89.

By Dr. Anup P. Tathod

19.	Li, J.; Spina, A.; Moulijn, J. A.; Makkee, M., Sorbitol dehydration into isosorbide in a molten salt hydrate medium. *Catalysis Science & Technology* **2013**, *3* (6), 1540-1546.

20.	Gohil, R. M., Properties and strain hardening character of polyethylene terephthalate containing Isosorbide. *Polymer Engineering & Science* **2009**, *49* (3), 544-553.

21.	Zhu, Y.; Durand, M.; Molinier, V.; Aubry, J.-M., Isosorbide as a novel polar head derived from renewable resources. Application to the design of short-chain amphiphiles with hydrotropic properties. *Green Chemistry* **2008**, *10* (5), 532-540.

22.	Davda, R. R.; Dumesic, J. A., Renewable hydrogen by aqueous-phase reforming of glucose. *Chemical Communications* **2004**, *0* (1), 36-37.

23.	He, L.; Chen, D., Hydrogen Production from Glucose and Sorbitol by Sorption-Enhanced Steam Reforming: Challenges and Promises. *ChemSusChem 5* (3), 587-595.

24.	Shawkat, H.; Westwood, M.-M.; Mortimer, A., Mannitol: a review of its clinical uses. *Continuing Education in Anaesthesia, Critical Care & Pain* **2012**.

25.	Mikkola, J. P.; Salmi, T., In-situ ultrasonic catalyst rejuvenation in three-phase hydrogenation of xylose. *Chemical Engineering Science* **1999**, *54* (10), 1583-1588.

26.	Mikkola, J. P.; Salmi, T., Three-phase catalytic hydrogenation of xylose to xylitol- prolonging the catalyst activity by means of on-line ultrasonic treatment. *Catalysis Today* **2001**, *64* (3-4), 271-277.

27.	Hildebrandt, G.; Lee, I.; Hodges, J., Oral mutans streptococci levels following use of a xylitol mouth rinse: a double-blind, randomized, controlled clinical trial. *Special Care in Dentistry* **2010**, *30* (2), 53-58.

28.	Lif Holgerson, P.; Stecksén-Blicks, C.; Sjöström, I.; Öberg, M.; Twetman, S., Xylitol Concentration in Saliva and Dental Plaque after Use of Various Xylitol-Containing Products. *Caries Research* **2006**, *40* (5), 393-397.

29.	Sano, H.; Nakashima, S.; Songpaisan, Y.; Phantumvanit, P., Effect of a xylitol and fluoride containing toothpaste on the remineralization of human enamel <I>in vitro</I>. *Journal of Oral Science* **2007**, *49* (1), 67-73.

30.	Makinen, K. K.; Bennett, C. A.; Hujoel, P. P.; Isokangas, P. J.; Isotupa, K. P.; Pape, H. R.; Makinen, P. L., Xylitol Chewing Gums and Caries Rates: A 40-month Cohort Study. *Journal of Dental Research* **1995**, *74* (12), 1904-1913.

31.	Uhari, M.; Kontiokari, T.; Koskela, M.; Niemela, M., *Xylitol chewing gum in prevention of acute otitis media: double blind randomised trial*. 1996; Vol. 313, p 1180-1183.

32.	da Silva, S. S. r.; Chandel, A. K.; de Cássia Lacerda Brambilla Rodrigues, R.; Canettieri, E.; Martinez, E.; Canilha, L.; Solenzal, A.; de Almeida e Silva, J. o., Statistical Approaches for the Optimization of Parameters for Biotechnological Production of Xylitol. In *D-Xylitol*, Springer Berlin Heidelberg: 2012; pp 133-160.

33.	de Kalbermatten, N.; Ravussin, E.; Maeder, E.; Geser, C.; Jéquier, E.; Felber, J. P., Comparison of glucose, fructose, sorbitol, and xylitol utilization in humans during insulin suppression. *Metabolism* **1980**, *29* (1), 62-67.

34.	Natah, S. S.; Hussien, K. R.; Tuominen, J. A.; Koivisto, V. A., Metabolic response to lactitol and xylitol in healthy men. *The American Journal of Clinical Nutrition* **1997**, *65* (4), 947-50.

35.	Chen, X.; Jiang, Z.-H.; Chen, S.; Qin, W., Microbial and Bioconversion Production of D-xylitol and Its Detection and Application. *International Journal of Biological Sciences* **2010**, *6* (7), 834-844.

36.	Sani, Y. M.; Daud, W. M. A. W.; Abdul Aziz, A. R., Activity of solid acid catalysts for biodiesel production: A critical review. *Applied Catalysis A: General* **2014**, *470* (0), 140-161.

37.	M. Van Rhijn, W.; E. De Vos, D.; F. Sels, B.; D. Bossaert, W., Sulfonic acid functionalised ordered mesoporous materials as catalysts for condensation and esterification reactions. *Chemical Communications* **1998**, (3), 317-318.

38.	Sharma, Y. C.; Singh, B.; Korstad, J., Advancements in solid acid catalysts for ecofriendly and economically viable synthesis of biodiesel. *Biofuels, Bioproducts and Biorefining* **2010**, *5* (1), 69-92.

39.   Gunnewegh, E. A.; Gopie, S. S.; van Bekkum, H., MCM-41 type molecular sieves as catalysts for the Friedel-Crafts acylation of 2-methoxynaphthalene. *Journal of Molecular Catalysis A: Chemical* **1996**, *106* (1â€"2), 151-158.
40.   Richard Kloetstra, K.; C. Jansen, J., Mesoporous material containing framework tectosilicate by pore-wall recrystallization. *Chemical Communications* **1997**, (23), 2281-2282.
41.   Trueba, M.; Trasatti, S. P., γ-Alumina as a Support for Catalysts: A Review of Fundamental Aspects. *European Journal of Inorganic Chemistry* **2005**, *2005* (17), 3393-3403.
42.   Zhou, R. S.; Snyder, R. L., Structures and transformation mechanisms of the [eta], [gamma] and [theta] transition aluminas. *Acta Crystallographica Section B* **1991**, *47* (5), 617-630.
43.   Tsyganenko, A. A.; Mardilovich, P. P., Structure of alumina surfaces. *Journal of the Chemical Society, Faraday Transactions* **1996**, *92* (23), 4843-4852.
44.   Ionescu, A.; Allouche, A.; Aycard, J.-P.; Rajzmann, M.; Hutschka, F. o., Study of Î³-Alumina Surface Reactivity:â€‰ Adsorption of Water and Hydrogen Sulfide on Octahedral Aluminum Sites. *The Journal of Physical Chemistry B* **2002**, *106* (36), 9359-9366.
45.   Coster, D.; Fripiat, J. J., Memory effects in gel-solid transformations: coordinately unsaturated aluminum sites in nanosized aluminas. *Chemistry of Materials* **1993**, *5* (9), 1204-1210.
46.   Hattori, H., Heterogeneous Basic Catalysis. *Chemical Reviews* **1995**, *95* (3), 537-558.
47.   Centi, G.; Perathoner, S., Catalysis by layered materials: A review. *Microporous and Mesoporous Materials* **2008**, *107* (1-2), 3-15.
48.   Vaccari, A., Preparation and catalytic properties of cationic and anionic clays. *Catalysis Today* **1998**, *41* (1â€"3), 53-71.
49.   Tichit, D.; Coq, B., Catalysis by Hydrotalcites and Related Materials. *CATTECH* **2003**, *7* (6), 206-217.
50.   Corma, A.; Iborra, S.; Bruce, C. G.; Helmut, K. z., Optimization of Alkaline Earth Metal Oxide and Hydroxide Catalysts for Base-Catalyzed Reactions. In *Advances in Catalysis*, Academic Press: 2006; Vol. Volume 49, pp 239-302.
51.   Figueras, F. o., Base Catalysis in the Synthesis of Fine Chemicals. *Topics in Catalysis* **2004**, *29* (3-4), 189-196.
52.   Debecker, D. P.; Gaigneaux, E. M.; Busca, G., Exploring, Tuning, and Exploiting the Basicity of Hydrotalcites for Applications in Heterogeneous Catalysis. *Chemistry – A European Journal* **2009**, *15* (16), 3920-3935.
53.   Liu, Y.; Lotero, E.; Goodwin Jr, J. G.; Mo, X., Transesterification of poultry fat with methanol using Mgâ€"Al hydrotalcite derived catalysts. *Applied Catalysis A: General* **2007**, *331* (0), 138-148.
54.   Li, F.; Jiang, X.; Evans, D.; Duan, X., Structure and Basicity of Mesoporous Materials from Mg/Al/In Layered Double Hydroxides Prepared by Separate Nucleation and Aging Steps Method. *Journal of Porous Materials* **2005**, *12* (1), 55-63.
55.   Sharma, S. K.; Parikh, P. A.; Jasra, R. V., Solvent free aldol condensation of propanal to 2-methylpentenal using solid base catalysts. *Journal of Molecular Catalysis A: Chemical* **2007**, *278* (1-2), 135-144.
56.   Roelofs, J. C. A. A.; Lensveld, D. J.; van Dillen, A. J.; de Jong, K. P., On the Structure of Activated Hydrotalcites as Solid Base Catalysts for Liquid-Phase Aldol Condensation. *Journal of Catalysis* **2001**, *203* (1), 184-191.
57.   Xie, W.; Peng, H.; Chen, L., Calcined Mgâ€"Al hydrotalcites as solid base catalysts for methanolysis of soybean oil. *Journal of Molecular Catalysis A: Chemical* **2006**, *246* (1-2), 24-32.
58.   KuÅ›trowski, P.; Chmielarz, L.; Rafalska-Å‚asocha, A.; Dudek, B.; Pattek-Janczyk, A.; Dziembaj, R., Catalytic reduction of N2O by ethylbenzene over novel hydrotalcite-derived Mgâ€"Crâ€"Feâ€"O as an alternative route for simultaneous N2O abatement and styrene production. *Catalysis Communications* **2006**, *7* (12), 1047-1052.
59.   Tichit, D.; GÃ©rardin, C.; Durand, R.; Coq, B., Layered double hydroxides: precursors for multifunctional catalysts. *Topics in Catalysis* **2006**, *39* (1-2), 89-96.

By Dr. Anup P. Tathod

60.   Cui, P.; Zhao, G.; Ren, H.; Huang, J.; Zhang, S., Ionic liquid enhanced alkylation of iso-butane and 1-butene. *Catalysis Today 200* (0), 30-35.

61.   Kuwahara, J.; Ikari, R.; Murata, K.; Nakamura, N.; Ohno, H., Electrocatalytic reduction of oxygen by bilirubin oxidase in hydrophobic ionic liquids containing a small quantity of water. *Catalysis Today 200* (0), 49-53.

62.   Diao, Y.; Li, J.; Wang, L.; Yang, P.; Yan, R.; Jiang, L.; Zhang, H.; Zhang, S., Ethylene hydroformylation in imidazolium-based ionic liquids catalyzed by rhodiumâ€"phosphine complexes. *Catalysis Today 200* (0), 54-62.

63.   Alonso, D. M.; Wettstein, S. G.; Dumesic, J. A., Bimetallic catalysts for upgrading of biomass to fuels and chemicals. *Chemical Society Reviews* **2012**, *41* (24), 8075-8098.

64.   Granstrom, T.; Izumori, K.; Leisola, M., A rare sugar xylitol. Part II: biotechnological production and future applications of xylitol. *Applied Microbiology and Biotechnology* **2007**, *74* (2), 273-276.

65.   Lichtenthaler, F. W., Unsaturated O- and N-heterocycles from carbohydrate feedstocks. *Acc Chem Res* **2002**, *35* (9), 728-37.

66.   Yadav, M.; Mishra, D. K.; Hwang, J. S., Catalytic hydrogenation of xylose to xylitol using ruthenium catalyst on NiO modified TiO2 support. *Applied Catalysis A: General* **2012**, *425-426* (0), 110-116.

67.   Zhang, J.; Geng, A.; Yao, C.; Lu, Y.; Li, Q., Xylitol production from d-xylose and horticultural waste hemicellulosic hydrolysate by a new isolate of Candida athensensis SB18. *Bioresource Technology* **2012**, *105* (0), 134-141.

68.   Yoshitake, J.; Ohiwa, H.; Shimamura, M.; Imai, T., Production of Polyalcohol by a <i>Corynebacterium</i> sp

Part I. Production of Pentitol from Aldopentose. *Agricultural and Biological Chemistry* **1971**, *35* (6), 905-911.

69.   Yoshitake, J.; Ishizaki, H.; Shimamura, M.; Imai, T., Xylitol Production by an <i>Enterobacter</i> Species. *Agricultural and Biological Chemistry* **1973**, *37* (10), 2261-2267.

70.   Yoshitake, J.; Shimamura, M.; Ishizaki, H.; Irie, Y., Xylitol Production by <i>Enterobacter liquefaciens</i>. *Agricultural and Biological Chemistry* **1976**, *40* (8), 1493-1503.

71.   Izumori, K.; Tuzaki, K., Production of xylitol from D-xylulose by Mycobacterium smegmatis. *Journal of Fermentation Technology* **1988**, *66* (1), 33-36.

72.   Dahiya, J. S., Xylitol production by Petromyces albertensis grown on medium containing D-xylose. *Canadian Journal of Microbiology* **1991**, *37* (1), 14-18.

73.   Oh, D. K.; Kim, S. Y.; Kim, J. H., Increase of xylitol production rate by controlling redox potential in Candida parapsilosis. *Biotechnology and Bioengineering* **1998**, *58* (4), 440-444.

74.   Baek, H.; Song, K.-H.; Park, S.-M.; Kim, S.-Y.; Hyun, H.-H., Role of glucose in the bioconversion of fructose into mannitol by Candida magnoliae. *Biotechnology Letters* **2003**, *25* (10), 761-765.

75.   Erzinger, G. S.; Silveira, M. M. d.; Costa, J. P. C. L. d.; Vitolo, M.; Jonas, R., Activity of glucose-fructose oxidoreductase in fresh and permeabilised cells of Zymomonas mobilis grown in different glucose concentrations. *Brazilian Journal of Microbiology* **2003**, *34*, 329-333.

76.   Saha, B. C.; Bothast, R. J., Production of L-arabitol from L-arabinose by Candida entomaea and Pichia guilliermondii. *Applied Microbiology and Biotechnology* **1996**, *45* (3), 299-306.

77.   Mikkola, J. P.; Salmi, T., Three-phase catalytic hydrogenation of xylose to xylitol â€" prolonging the catalyst activity by means of on-line ultrasonic treatment. *Catalysis Today* **2001**, *64* (3-4), 271-277.

78.   Herskowitz, M., Modelling of a trickle-bed reactor-the hydrogenation of xylose to xylitol. *Chemical Engineering Science* **1985**, *40* (7), 1309-1311.

79.   van Gorp, K.; Boerman, E.; Cavenaghi, C. V.; Berben, P. H., Catalytic hydrogenation of fine chemicals: sorbitol production. *Catalysis Today* **1999**, *52* (2-3), 349-361.

By Dr. Anup P. Tathod

80.   Gallezot, P.; Nicolaus, N.; FlÃ¨che, G.; Fuertes, P.; Perrard, A., Glucose Hydrogenation on Ruthenium Catalysts in a Trickle-Bed Reactor. *Journal of Catalysis* **1998**, *180* (1), 51-55.

81.   Mikkola, J.-P.; SjÃ¶holm, R.; Salmi, T.; MÃ¤ki-Arvela, P. i., Xylose hydrogenation: kinetic and NMR studies of the reaction mechanisms. *Catalysis Today* **1999**, *48* (1-4), 73-81.

82.   Schimpf, S.; Louis, C.; Claus, P., Ni/SiO2 catalysts prepared with ethylenediamine nickel precursors: Influence of the pretreatment on the catalytic properties in glucose hydrogenation. *Applied Catalysis A: General* **2007**, *318* (0), 45-53.

83.   Lux, C.; Wollenhaupt, M.; Sarpe, C.; Baumert, T., Photoelectron Circular Dichroism of Bicyclic Ketones from Multiphoton Ionization with Femtosecond Laser Pulses. *ChemPhysChem 16* (1), 7-7.

84.   Jeon, W. Y.; Yoon, B. H.; Shim, W. Y.; Kim, J. H., Reduction of arabitol in xylitol production by changed substrate preference of xylose reductase in Candida tropicalis. *Journal of Biotechnology* **2010**, *150, Supplement* (0), 345-346.

85.   Ling, H.; Cheng, K.; Ge, J.; Ping, W., Statistical optimization of xylitol production from corncob hemicellulose hydrolysate by Candida tropicalis HDY-02. *New Biotechnology* **2011**, *28* (6), 673-678.

86.   Kusserow, B.; Schimpf, S.; Claus, P., Hydrogenation of Glucose to Sorbitol over Nickel and Ruthenium Catalysts. *Advanced Synthesis & Catalysis* **2003**, *345* (1-2), 289-299.

87.   Hoffer, B. W.; Crezee, E.; Mooijman, P. R. M.; van Langeveld, A. D.; Kapteijn, F.; Moulijn, J. A., Carbon supported Ru catalysts as promising alternative for Raney-type Ni in the selective hydrogenation of d-glucose. *Catalysis Today* **2003**, *79-80* (0), 35-41.

88.   Arena, B. J., Deactivation of ruthenium catalysts in continuous glucose hydrogenation. *Applied Catalysis A: General* **1992**, *87* (2), 219-229.

89.   Kuusisto, J.; Tokarev, A. V.; Murzina, E. V.; Roslund, M. U.; Mikkola, J.-P.; Murzin, D. Y.; Salmi, T., From renewable raw materials to high value-added fine chemicalsâ€"Catalytic hydrogenation and oxidation of d-lactose. *Catalysis Today* **2007**, *121* (1-2), 92-99.

90.   Perrard, A.; Gallezot, P.; Joly, J.-P.; Durand, R.; Baljou, C. d.; Coq, B.; Trens, P., Highly efficient metal catalysts supported on activated carbon cloths: A catalytic application for the hydrogenation of d-glucose to d-sorbitol. *Applied Catalysis A: General* **2007**, *331* (0), 100-104.

91.   Fukuoka, A.; Dhepe, P. L., Catalytic Conversion of Cellulose into Sugar Alcohols. *Angewandte Chemie International Edition* **2006**, *45* (31), 5161-5163.

92.   Geboers, J.; Van de Vyver, S.; Carpentier, K.; Jacobs, P.; Sels, B., Efficient hydrolytic hydrogenation of cellulose in the presence of Ru-loaded zeolites and trace amounts of mineral acid. *Chemical Communications* **2011**, *47* (19), 5590-5592.

93.   Ding, L.-N.; Wang, A.-Q.; Zheng, M.-Y.; Zhang, T., Selective Transformation of Cellulose into Sorbitol by Using a Bifunctional Nickel Phosphide Catalyst. *ChemSusChem 3* (7), 818-821.

94.   Kobayashi, H.; Ito, Y.; Komanoya, T.; Hosaka, Y.; Dhepe, P. L.; Kasai, K.; Hara, K.; Fukuoka, A., Synthesis of sugar alcohols by hydrolytic hydrogenation of cellulose over supported metal catalysts. *Green Chemistry* **2011**, *13* (2), 326-333.

95.   Shrotri, A.; Tanksale, A.; Beltramini, J. N.; Gurav, H.; Chilukuri, S. V., Conversion of cellulose to polyols over promoted nickel catalysts. *Catalysis Science & Technology* **2012**, *2* (9), 1852-1858.

96.   Sharkov, V. I., Production of Polyhydric Alcohols from Wood Polysaccharides. *Angewandte Chemie International Edition in English* **1963**, *2* (8), 405-409.

97.   Luo, C.; Wang, S.; Liu, H., Cellulose Conversion into Polyols Catalyzed by Reversibly Formed Acids and Supported Ruthenium Clusters in Hot Water. *Angewandte Chemie International Edition* **2007**, *46* (40), 7636-7639.

98.   Deng, T.; Liu, H., Promoting effect of SnOx on selective conversion of cellulose to polyols over bimetallic Pt-SnOx/Al2O3 catalysts. *Green Chemistry* **2013**, *15* (1), 116-124.

99.   Felipe, M. G. A.; Vitolo, M.; Mancilha, I. M.; Silva, S. S., Fermentation of sugar cane bagasse hemicellulosic hydrolysate for xylitol production: Effect of pH. *Biomass and Bioenergy* **1997**, *13* (1-2), 11-14.

By Dr. Anup P. Tathod

100.   Zhang, J.; Geng, A.; Yao, C.; Lu, Y.; Li, Q., Xylitol production from d-xylose and horticultural waste hemicellulosic hydrolysate by a new isolate of Candida athensensis SB18. *Bioresource Technology* **2011,** *105* (0), 134-141.

101.   Canilha, L.; Carvalho, W.; Felipe, M. d. G. a. A.; Silva, J. o. B. d. A. e., Xylitol production from wheat straw hemicellulosic hydrolysate: hydrolysate detoxification and carbon source used for inoculum preparation. *Brazilian Journal of Microbiology* **2008,** *39,* 333-336.

102.   Nair, N. U.; Zhao, H., Selective reduction of xylose to xylitol from a mixture of hemicellulosic sugars. *Metab Eng* **2010,** *12* (5), 7-7.

103.   Carvalho, W.; Silva, S. S.; Vitolo, M.; Felipe, M. G.; Mancilha, I. M., Improvement in xylitol production from sugarcane bagasse hydrolysate achieved by the use of a repeated-batch immobilized cell system. *Z Naturforsch C* **2002,** *57* (1-2), 109-112.

104.   Yi, G.; Zhang, Y., One-Pot Selective Conversion of Hemicellulose (Xylan) to Xylitol under Mild Conditions. *ChemSusChem 5* (8), 1383-1387.

105.   Guha, S. K.; Kobayashi, H.; Hara, K.; Kikuchi, H.; Aritsuka, T.; Fukuoka, A., Hydrogenolysis of sugar beet fiber by supported metal catalyst. *Catalysis Communications* **2011,** *12* (11), 980-983.

106.   Kusema, B. T.; Faba, L.; Kumar, N.; MÃ¤ki-Arvela, P. i.; DÃaz, E.; OrdÃ³Ã±ez, S.; Salmi, T.; Murzin, D. Y., Hydrolytic hydrogenation of hemicellulose over metal modified mesoporous catalyst. *Catalysis Today* **2012,** *196* (1), 26-33.

107.   Faba, L.; Kusema, B. T.; Murzina, E. V.; Tokarev, A.; Kumar, N.; Smeds, A.; DÃaz, E.; OrdÃ³Ã±ez, S.; MÃ¤ki-Arvela, P. i.; WillfÃ¶r, S.; Salmi, T.; Murzin, D. Y., Hemicellulose hydrolysis and hydrolytic hydrogenation over proton- and metal modified beta zeolites. *Microporous and Mesoporous Materials* (0).

108.   Kim, Y.; Hendrickson, R.; Mosier, N.; Ladisch, M. R., Plug-Flow Reactor for Continuous Hydrolysis of Glucans and Xylans from Pretreated Corn Fiber. *Energy & Fuels* **2005,** *19* (5), 2189-2200.

109.   Jinesh, C. M.; Antonyraj, C. A.; Kannan, S., Allylbenzene isomerisation over as-synthesized MgAl and NiAl containing LDHs: Basicity-activity relationships. *Applied Clay Science* **2010,** *48* (1-2), 243-249.

110.   Dhepe, P. L.; Sahu, R., A solid-acid-based process for the conversion of hemicellulose. *Green Chemistry* **2010,** *12* (12), 2153-2156.

111.   Cara, P. D.; Pagliaro, M.; Elmekawy, A.; Brown, D. R.; Verschuren, P.; Shiju, N. R.; Rothenberg, G., Hemicellulose hydrolysis catalysed by solid acids. *Catalysis Science & Technology* **2013,** *3* (8), 2057-2061.

112.   Ormsby, R.; Kastner, J. R.; Miller, J., Hemicellulose hydrolysis using solid acid catalysts generated from biochar. *Catalysis Today* **2012,** *190* (1), 89-97.

113.   Sahu, R.; Dhepe, P. L., A One-Pot Method for the Selective Conversion of Hemicellulose from Crop Waste into C5 Sugars and Furfural by Using Solid Acid Catalysts. *ChemSusChem* **2012,** *5* (4), 751-761.

114.   Li, N.; Tompsett, G. A.; Zhang, T.; Shi, J.; Wyman, C. E.; Huber, G. W., Renewable gasoline from aqueous phase hydrodeoxygenation of aqueous sugar solutions prepared by hydrolysis of maple wood. *Green Chemistry 13* (1), 91-101.

115.   Abasaeed, A. E.; Lee, Y. Y., Inulin hydrolysis to fructose by a novel catalyst. *Chemical Engineering & Technology* **1995,** *18* (6), 440-444.

116.   Heinen, A. W.; Peters, J. A.; van Bekkum, H., The combined hydrolysis and hydrogenation of inulin catalyzed by bifunctional Ru/C. *Carbohydrate Research* **2001,** *330* (3), 381-390.

117.   Bhaumik, P.; Dhepe, P. L., Exceptionally high yields of furfural from assorted raw biomass over solid acids. *RSC Advances* **2014,** *4* (50), 26215-26221.

118.   Bhaumik, P.; Dhepe, P. L., Efficient, Stable, and Reusable Silicoaluminophosphate for the One-Pot Production of Furfural from Hemicellulose. *ACS Catalysis* **2013,** *3* (10), 2299-2303.

119.   Tathod, A. P.; Dhepe, P. L., Efficient method for the conversion of agricultural waste into sugar alcohols over supported bimetallic catalysts. *Bioresource Technology* **2014**, *178* (0), 36-44.

120.   Tathod, A. P.; Gazit, O. M., Fundamental Insights into the Nucleation and Growth of Mg–Al Layered Double Hydroxides Nanoparticles at Low Temperature. *Crystal Growth & Design* **2016**, *16* (12), 6709-6713.

121.   Tathod, A. P.; Dhepe, P. L., Towards efficient synthesis of sugar alcohols from mono- and poly-saccharides: Role of metals, supports & promoters. *Green Chemistry* **2014**.

122.   Tathod, A.; Kane, T.; Sanil, E. S.; Dhepe, P. L., Solid base supported metal catalysts for the oxidation and hydrogenation of sugars. *Journal of Molecular Catalysis A: Chemical* **2014**, *388-389* (0), 90-99.

123.   Tathod, A. P.; Dhepe, P. L., Elucidating the effect of solid base on the hydrogenation of C5 and C6 sugars over Pt–Sn bimetallic catalyst at room temperature. *Carbohydrate Research* **2021**, *505*, 108341.

Appendix- I: List of relevant publications by the author

1. Solid base supported metal catalysts for the oxidation and hydrogenation of sugars

   Anup Tathod, Tanushree Kane, E.S. Sanil, Paresh L. Dhepe∗ *J. Mol. Catal. A: Chem.*, 2014, 388–389, 90–99

2. Towards efficient synthesis of sugar alcohols from mono- and poly-saccharides: role of metals, supports & promoters

   Anup Tathod, Paresh Dhepe* *Green Chem.*, 2014, 16, 4944-4954

3. Efficient method for the conversion of agricultural waste into sugar

   alcohols over supported bimetallic catalysts

   Anup Tathod, Paresh Dhepe* *Bioresour. Technol.*, 2015, 178, 36-44

4. Fundamental Insights into the Nucleation and Growth of Mg-Al Layered Double Hydroxides Nanoparticles at Low Temperature

   Anup Tathod, Oz Gazit, *Crystal Growth & Design*, 2016, *16* (12), 6709–6713

5. Mediating interaction strength between nickel and zirconia using a mixed oxide nanosheets interlayer for methane dry reforming

   Anup Tathod, Naseem Hayek, Dina Shpasser, David Simakov and Oz M. Gazit

   Applied Catalysis B: Environmental, 2019, 249,106-115

6. Elucidating the effect of solid base on the hydrogenation of C5 and C6 sugars over Pt–Sn bimetallic catalyst at room temperature.

   Anup Tathod, Paresh Dhepe, *Carbohydrate Research,* 2021, *505*, 108341.

7. Patent (India): Production of sugar alcohols, xylitol and arabitol using bimetallic supported metal catalysts, Anup Tathod, Paresh Dhepe. IN 0017/DEL/2014.

Publisher: Eliva Press SRL

Email: info@elivapress.com

Eliva Press is an independent publishing house established for the publication and dissemination of academic works all over the world. Company provides high quality and professional service for all of our authors.

Our Services:
Free of charge, open-minded, eco-friendly, innovational.

-Free standard publishing services (manuscript review, step-by-step book preparation, publication, distribution, and marketing).
-No financial risk. The author is not obliged to pay any hidden fees for publication.
-Editors. Dedicated editors will assist step by step through the projects.
-Money paid to the author for every book sold. Up to 50% royalties guaranteed.
-ISBN (International Standard Book Number). We assign a unique ISBN to every Eliva Press book.
-Digital archive storage. Books will be available online for a long time. We don't need to have a stock of our titles. No unsold copies. Eliva Press uses environment friendly print on demand technology that limits the needs of publishing business. We care about environment and share these principles with our customers.
-Cover design. Cover art is designed by a professional designer.
-Worldwide distribution. We continue expanding our distribution channels to make sure that all readers have access to our books.

www.elivapress.com

www.ingramcontent.com/pod-product-compliance
Lightning Source LLC
Chambersburg PA
CBHW051251170526
45165CB00004B/1663